Multiple Stresses in Ecosystems

Multiple Stresses in Ecosystems

Edited by

Joseph J. Cech, Jr.
Department of Wildlife, Fish, and Conservation Biology
University of California – Davis

Barry W. Wilson
Departments of Avian Sciences and Environmental Toxicology
University of California – Davis

Donald G. Crosby
Department of Toxicology
University of California – Davis

Lewis Publishers

Boca Raton Boston London New York Washington, D.C.

Library of Congress Cataloging-in-Publication Data

Multiple stresses in ecosystems / edited by Joseph J. Cech, Jr., Barry W. Wilson, Donald G. Crosby.
p. cm.
Resulted from an international conference held Oct. 14–15, 1993 and sponsored by the University of California Toxic Substances Program.
Includes bibliographical references and index.
ISBN 1-56670-309-3 (alk. paper)
1. Pollution—Environmental aspects—Congresses. 2. Nature—Effect of human beings on—Congresses. 3. Ecology—Congresses.
I. Cech, Joseph J. II. Wilson, Barry W., 1931– . III. Crosby, Donald G.
QH545.A1M85 1998
577.27—dc21

97-49109
CIP

Lewis Publishers is an imprint of CRC Press LLC

International Standard Book Number 1-56670-309-3
Library of Congress Card Number 97-49109
Printed in the United States of America 1 2 3 4 5 6 7 8 9 0
Printed on acid-free paper

The Editors

Joseph J. Cech, Jr., is Professor of Fish Biology/Physiology at the University of California, Davis. He earned degrees in zoology from the University of Wisconsin, Madison (B.S.) and the University of Texas, Austin (M.A., Ph.D.). After post-doctoral research experience at the Research Institute of the Gulf of Maine, he joined the UC Davis faculty in 1975. He regularly teaches Physiology of Fishes, Physiological Ecology, and Field Studies in Fish Biology courses. With his students and postdoctoral researchers, he is active in research on the functional responses of fishes to their environments (including contaminated ones) and is the author (or co-author) of over 90 research publications, including several books. For example, he co-authored (with Dr. Peter B. Moyle) *Fishes: An Introduction to Ichthyology*, now in its third edition.

Barry W. Wilson is an ecotoxicologist/neurotoxicologist and Professor of Avian Sciences and Environmental Toxicology at the University of California, Davis. Wilson was educated in liberal arts at the University of Chicago, in biology at the Illinois Institute of Technology, and in zoology at UCLA, where he obtained his Ph.D. degree. His biomedical research includes work on neuromuscle abnormalities, muscular dystrophy, acute pesticide exposures, and polyneuropathies. Recently he and his colleagues have been developing biomarkers of exposure and effect for wildlife, including hepatocyte, muscle, and nerve cell cultures, blood cholinesterase assays, and fecal testosterone indices of the reproductive state.

Donald G. Crosby is Professor Emeritus of Environmental Toxicology at the University of California, Davis. With a Ph.D. in chemistry from Cal Tech and almost 8 years' research at Union Carbide, he joined UC Davis in 1961 to start its Environmental Toxicology Department. He was a founding member of the American Chemical Society Division of Pesticide Chemistry (now Agrochemicals), SETAC, and the UC Davis Ecotoxicology Program, and still serves on the editorial boards of *Ecotoxicology and Environmental Safety* and *Reviews of Environmental Contamination and Toxicology.* He is author of a new book, *Environmental Toxicology and Chemistry.*

Contributors

S. Marshall Adams, Ph.D.
Environmental Sciences Division
Oak Ridge National Laboratory
Oak Ridge, Tennessee

Daniel W. Anderson, Ph.D.
Department of Wildlife, Fish, and Conservation Biology
University of California, Davis
Davis, California

Joseph J. Cech, Jr., Ph.D.
Department of Wildlife, Fish, and Conservation Biology
University of California, Davis
Davis, California

Gary N. Cherr, Ph.D.
University of California, Davis
Department of Environmental Toxicology and
Bodega Marine Laboratory
Bodega Bay, California

Donald G. Crosby, Ph.D.
Department of Environmental Toxicology
University of California, Davis
Davis, California

Anne Fairbrother, D.V.M.,Ph.D.
ecological planning and toxicology, inc.
Corvallis, Oregon

Shirley J. Gee, M.S.
Departments of Entomology and Environmental Toxicology
University of California, Davis
Davis, California

Edward D. Goldberg, Ph.D.
Scripps Institution of Oceanography
Marine Research Division
University of California
La Jolla, California

Charles R. Goldman, Ph.D.
Division of Environmental Studies
University of California, Davis
Davis, California

K. D. Ham, Ph.D.
Environmental Sciences Division
Oak Ridge National Laboratory
Oak Ridge, Tennessee

Bruce D. Hammock, Ph.D.
Departments of Entomology and Environmental Toxicology
University of California, Davis
Davis, California

Adam S. Harris, B.S.
Departments of Entomology and Environmental Toxicology
University of California, Davis
Davis, California

Alan G. Heath, Ph.D.
Department of Biology
Virginia Polytechnic Institute and State University
Blacksburg, Virginia

Sabine B. Kreissig, Ph.D.
Departments of Entomology and Environmental Toxicology
University of California, Davis
Davis, California

Bill L. Lasley, Ph.D.
Institute of Toxicology and
Environmental Health
University of California, Davis
Davis, California

R. F. LeHew, M.S.
Environmental Sciences Division
Oak Ridge National Laboratory
Oak Ridge, Tennessee

Simon A. Levin, Ph.D.
Department of Ecology and
Evolutionary Biology
Princeton University
Princeton, New Jersey

Thomas E. McKone, Ph.D.
University of California
School of Public Health
and Lawrence Berkeley Laboratory
Berkeley, California

Pierre Mineau, Ph.D.
National Wildlife Research Centre
Canadian Wildlife Service
Hull, Québec, Canada

James N. Seiber, Ph.D.
University Center for Environmental
Sciences and Engineering
Department of Environmental and
Resource Sciences
University of Nevada, Reno
Reno, Nevada

George E. Taylor, Jr., Ph.D.
College of Agriculture
University of Nevada, Reno
Reno, Nevada

Ronald S. Tjeerdema, Ph.D.
Department of Chemistry and
Biochemistry
University of California, Santa Cruz
Santa Cruz, California

Ingrid Wengatz, Ph.D.
Departments of Entomology and
Environmental Toxicology
University of California, Davis
Davis, California

Barry W. Wilson, Ph.D.
Departments of Avian Sciences and
Environmental Toxicology
University of California, Davis
Davis, California

Monika Wortberg, Ph.D.
Departments of Entomology and
Environmental Toxicology
University of California, Davis
Davis, California

Preface

The disciplinary areas of ecology and toxicology have made significant advances toward common goals of better detection, evaluation, and understanding of the complex interactions of toxicants in environmental settings. More than ever, we can piece together the physical, chemical, and biological influences of these substances and their effects on resident and migratory biota. We can often separate chronic from acute effects and follow substances through food webs, investigate processes at various levels of organization: from molecules and cells through tissues, organs, and organisms, to populations and communities, and are beginning to examine effects of more than one substance or impact on ecosystems. We are also assessing the health of ecosystems relevant to these impacts, and the risk to nonhuman and human inhabitants. The environmental impact review process and superfund site concerns are more closely linking the work of scientists in agencies, consulting firms, and universities. This volume is an attempt to pull together information from appropriate experts to assess: (1) where we are today in these relevant fields, and (2) which tools will be helpful in designing tomorrow's studies.

The volume is divided into three sections: Impact of Multiple Stresses on Ecosystems, Establishing the Health of Ecosystems, and Future Methods in Ecotoxicology. It resulted from an international conference on this topic held October 14–15, 1993. The conference was sponsored by the University of California Toxic Substances Program (Dr. Jerry Last, Director), especially through its Ecotoxicology Program (directed at that time by Dr. Daniel Anderson, now directed by Dr. David Hinton). The UC Ecotoxicology Program is a UC system-wide program that offers support, primarily in the form of traineeships, to students pursuing graduate degrees in areas such as ecology, pharmacology and toxicology, engineering, and others which relate to ecotoxicological investigations. The Davis Campus is currently the leader in the UC ecotoxicology thrust. The conference was also sponsored by the U.S. Environmental Protection Agency through its support of the Center for Ecological Health Research (Dr. Dennis Rolston, Director) located on the Davis campus.

Many individuals contributed to the success of the conference. Its program, leading to the chapters in this volume, was constructed by program committee members Drs. Daniel Anderson, Donald Crosby, and Thomas McKone. The co-chairs were Drs. Joseph Cech and Barry Wilson. Conference-related correspondence was handled by Ms. Marjorie Kirkman and Ms. Brenda Nakamoto of the UC Davis Department of Wildlife, Fish, and Conservation Biology. Ms. Evett Stranghellini-Kilmartin (UC Ecotoxicology Program) and Ms. Cheryl Smith (Center for Ecological Health Research) assisted with conference organization, facilities, publicity, and registration. Graduate student Monica Choi worked with Ms. Smith on the poster displays; and Nancy Autumn, Donna Bartkowiak, Seth Coleman, Steve Detwiler, Ruth Ann Elbert, Andrea Erikson, Adam Harris, Xiaoping Li, and Eric Mielbrecht assisted with registration, media projection, transcription, and transportation needs. Manuscript preparation was facilitated by Ms. Stranghellini-Kilmartin, Ms. Laura Brink, Ms. Kirkman, and Mr. Eric Paulovich (UC Davis Department of Wildlife, Fish, and Conservation Biology). Mr. Paulovich also facilitated the final editing and the transport of manuscript drafts among the publisher/reviewers, authors, and editors.

Joseph J. Cech, Jr.
Barry W. Wilson
Donald G. Crosby

Table of Contents

Part One

Impact of Multiple Stresses on Ecosystems

1 Evaluation and Impact of Multiple Stressors on Ecosystems: Four Classic Case Histories

Daniel W. Anderson

INTRODUCTION

"You can create excitement, you can do wonderful promotion and get all kinds of press... But if you don't deliver the goods, people will eventually catch on."

Trump (1987)

I am taking this quotation out of its intended context, and I have no particular feelings, one way or another, about how to conduct a business; but what Donald Trump says here pertains well to the field of ecotoxicology. Although wholly based on basic biology at all levels of approach or entry, ecotoxicology is ultimately an applied field — society expects results, because after all, most aspects of ecotoxicology ultimately involve one or more imposed, additional environmental stresses from contamination or other human-associated stressors. Results, however, of remediation in ecotoxicology depend entirely on an accurate understanding of natural or unaffected processes and the interactions of these factors with the stresses imposed through natural variation in the environment. Although there are many pseudo-comparisons between ecotoxicology and medicine (outlined in several chapters in this book), like medicine perhaps, we are in a field that intends to truly intermix the basic and applied sciences. Perhaps their separation and distinction is even an artifact!

ECOTOXICOLOGY

According to Moriarty (1988), the term "ecotoxicology" was apparently coined by Truhaut (1977), who saw it as a branch of toxicology. We ecologists would like to see it as a branch of ecology; and as the field has matured, ecology has played an ever larger role at all levels of entry (Figure 1.1), and that seems to be where the field is evolving today, also with a strong foundation on the molecular, the physiological, and the physical sciences. Ramade (1987) defines ecotoxicology as: "... the science whose aim ... is to study ... environmental contamination by ... pollutants ...,

1-56670-309-3/98/$0.00+$.50

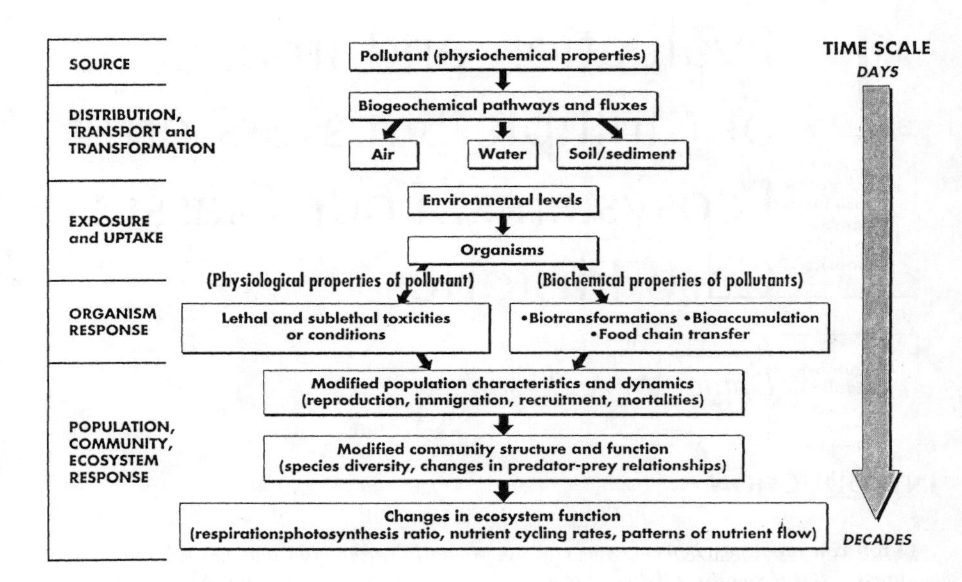

FIGURE 1.1 Diagram showing the many levels of approach in ecotoxicology and its integrative nature. (Modified from Connell, D.W. and Miller, G.J., *The Chemistry and Ecotoxicology of Pollution,* John Wiley & Sons, New York, 1987; Sheehan, P. J., in *Effects of Pollutants at the Ecosystem Level,* John Wiley & Sons, New York, 1984; Connell, D. W., *Ambio,* 16, 47, 1987; Reijnders, P. J. H., *Marine Mammal Sci.,* 4, 91, 1988.)

as well as … the mechanisms of their actions and effects … Ecotoxicology … has to acquire a profound knowledge of the fundamental concepts of ecology as well as … the physiological mechanisms governing the action of pollutants." Moriarty (1988) defines it as "effects on ecosystems." Levin et al. (1984, 1989) state that ecotoxicology is: "… the science that deals with ecosystem-level effects of toxic substances …" The University of California, Davis, Program in Ecotoxicology is based on Butler (1978): "Ecotoxicology is concerned with the toxic effects of chemical and physical agents on living organisms, especially on populations and communities within defined ecosystems; it includes the transfer pathways of those agents and their interactions within the environment …" Ecotoxicologists enter and study biological systems at many levels (Figure 1.1) and often at many levels simultaneously. Never before has the need for a team approach with specialists from many areas (Ramade 1987 calls ecotoxicology an "emerging specialist science") been more apparent in applied ecology. Depledge (1993) stated that "… in the rush to achieve effective environmental protection, we should all remember that effective test procedures and predictive models must be based on the findings of high-quality basic research."

So now, we have a perception of what ecotoxicology is: it is new, but not really; it is a science of itself, but it is not. Yet it attempts to be integrative, and it attempts to address in nontraditional ways, environmental and ecological problems that affect ecological systems and biodiversity, and in the context of interactions with "normal"

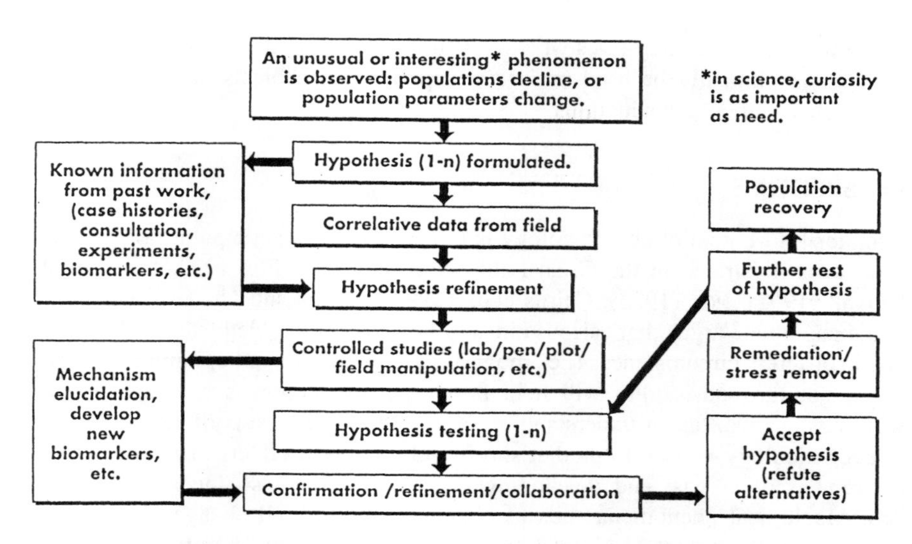

FIGURE 1.2 An idealized, stepwise reactive scientific determination of a negative toxicant-population interaction and its remediation, as it has generally occurred (or should have occurred) in most cases in the past. This diagram represents the usual pathway that has been followed when toxicant-related avian population declines occurred, but it applies in general to traditional ecotoxicological problem solving.

stressors, additional stressors not directly related to contaminants, and stressors associated with expected normal variability. Yet, ecotoxicology seems to offer few simple solutions to ecosystem management problems through pure technology. Its solutions are always complex because of the complexity of environmental "etiology" (Fox 1993). It is a science of warnings from small-scale systems that need to be heeded on a larger scale. Its ultimate goals are prevention rather than remediation.

THE REACTIVE APPROACH IN ECOTOXICOLOGY

> *"The central significance of [the] DDE affair for overall resource policy... is that there exists no effective mechanism for preventing such resource degradations by toxic chemicals now or in the future."*
>
> Keith (1969)

Almost all case-histories in ecotoxicological problem definition and then solution have been characterized by a reactive approach (Figure 1.2). We have always looked at problems after they are shown to already exist. We formulate hypotheses, test them in the field, the pen, and the laboratory. Problems are defined and solutions proposed — decision-makers are supposed to listen and do something about it. A further test of our original hypothesis comes after the contaminant/pollutant stress is removed (Figure 1.2). Fortunately, all biological systems are resilient to some degree, and this resiliency has often kept society out of serious trouble. But eventually we have to ask ourselves whether natural or even production-oriented ecosystems can continually be used as *in situ* experimental subjects or test systems in future environmental evaluations.

PERSPECTIVE

A major future goal of ecotoxicology is to become more anticipatory or proactive, such as is illustrated in the Great Lakes ecosystem (see Fox 1993, for example). Peakall (1992), Fox (1993), Cairns et al. (1993), Cairns and Niederlehner (1995), and many others argue that various kinds of biomarkers sampled from various levels and in different circumstances (i.e., monitoring the exposure or beginnings of impairment, not the end-results) will help provide ecotoxicologists a series of "early warnings" as well as an understanding of specific mechanisms of effect. The goal in ecotoxicology will be to predict short- and long-term effects of toxicants in the future, based on past and developing experiences, but also incorporating other ecotoxicological phenomena such as chemical transfer and change, the predicted chemical movement into and out of the affected ecosystems, monitoring bioindicators that signal these transfers, interpreting these dynamics in a well-grounded knowledge of the modifying effects of natural variations (including the considerations of added stressors), and the all-important policy reaction that allows remediation of the situations. It is hoped these actions can be accomplished before regional and large-scale system changes occur, such as we have experienced all too often in the past — or even before any serious or unwanted local changes might occur. Thus, the studies of the effects of toxicants on individuals, populations, communities, and systems must continue. There is an even greater need for data and models on how populations and systems operate under normal, varying environmental conditions, and multiple-stressors (Hart 1990).

I think that the best chances we have for "heading off" serious, future environmental problems in a more anticipatory fashion involve: (1) the use of biomarkers of sublethal and sub-effective effects as predictors or early-warning devices, such as is proposed by Fox (1993); (2) use of models that can look at the presence of various toxicants (this will require more and more sophisticated analytical chemistry as well as biomarkers) in parts of a system and predict when and if this presence is likely to become a problem, and which fully consider expected environmental variability and added potential stressors; and (3) continued and improved ability to respond to situations that "sneak up" on us as before, but with a more proactive approach than had previously developed mostly as a reactive response in ecotoxicology.

The next four chapters are case histories from four different systems which illustrate the depiction of problems, the evolution and refinement of the remedial and diagnostic approach, and updated results on the interpretation and significance of improvements in each situation. I consider these cases as typical, but "classic," cases of the ecotoxicological approach developed in the first 40 years of ecotoxicology, where lessons can be learned and such groundwork can lead to new ideas and synthesis for future endeavors.

REFERENCES

Butler, G. C., Ed., *Principles of Ecotoxicology (SCOPE)*, John Wiley & Sons, New York, 1978.

Cairns, J., Jr., McCormick, P. V., and Niederlehner, B. R., A proposed framework for developing indicators of ecosystem health, *Hydrobiologia*, 263, 1, 1993.

Cairns, J., Jr., and Niederlehner, B. R., Predictive ecotoxicology, in *Handbook of Ecotoxicology*, Hoffman, D. J., Rattner, B. A., Burton, G. A., and Cairns, J., Jr., Eds., CRC Press, Boca Raton, 1995, pp. 667-680.

Connell, D. W., Ecotoxicology — a framework for investigations of hazardous chemicals in the environment, *Ambio*, 16, 47, 1987.

Connell, D. W., and Miller, G. J., *The Chemistry and Ecotoxicology of Pollution*, John Wiley & Sons, New York, 1984.

Depledge, M. H., Ecotoxicology: a science or a management tool?, *Ambio*, 22, 51, 1993.

Fox, G. A., What have biomarkers told us about the effects of contaminants on the health of fish-eating birds in the Great Lakes? The theory and a literature review, *J. Great Lakes Res.*, 19, 722, 1993.

Hart, A. D. M., The assessment of pesticide hazards to birds: the problem of variable effects, *Ibis*, 132, 192, 1990.

Keith, J. A., The DDE affair, *Canadian Field-Naturalist*, 83, 89, 1969.

Levin, S. A., Kimball, K. D., McDowell, W. H., and Kimball, S. F., New perspectives in ecotoxicology, *Environ. Manage.*, 8, 375, 1984.

Levin, S. A., Hartwell, M. A., Kelly, J. R., and Kimball, K. D., *Ecotoxicology: Problems and Approaches*, Springer-Verlag, New York, 1989.

Moriarty, F, *Ecotoxicology: The Study of Pollutants in Ecosystems*, Academic Press, New York, 1988.

Peakall, D., *Animal Biomarkers as Pollution Indicators*, Chapman & Hall, New York, 1992.

Ramade, F., *Ecotoxicology*, John Wiley & Sons, New York, 1987.

Reijnders, P. J. H., Ecotoxicological perspectives in marine mammalogy: research principles and goals for a conservation policy, *Marine Mammal Sci.*, 4, 91, 1988.

Sheehan, P. J., Effects on community and ecosystem structure and dynamics, in *Effects of Pollutants at the Ecosystem Level*, Sheehan, P. J., Miller, D. R., Butler, G. C., and Bourdeau, P., Eds., John Wiley & Sons, New York, 1984, pp. 101-145.

Truhaut, R., Ecotoxicology: objectives, principles and perspectives, *Ecotox. Environ. Safety*, 1, 151, 1977.

Trump, D. J. (with Schwartz, T.), *Trump: The Art of the Deal*, Random House, New York, 1987.

2 Extrapolation and Scaling in Ecotoxicology

Simon A. Levin

Ecotoxicology is about the fate, transport, and effects of chemicals in the environment. The principal challenges are description and prediction, challenges that become much harder when the sources are diffuse, as they are for most regional and global ecological problems. Furthermore, the theory for dealing with fate and transport is much better developed than the theory for dealing with effects. As substances become distributed through the food chain, there will be effects upon organisms; those effects will feed back to modify not only effects on other organisms, but also the flows of material.

The fundamental problem is one of extrapolation: from laboratory to field and from the effects on single individuals to the effects on communities and ecosystems. In order to achieve prediction, one needs a mechanistic approach that is tied to individual behavior. Therefore, the models that I will discuss are ones that begin from effects on individuals and scale up to effects on whole systems.

In dealing, for example, with air pollution effects on forest plants, one must extrapolate from information on the individual responses of plants in controlled environments to field situations, and ultimately to whole communities in which nonlinear interactions become important. In the usual models of vegetational responses to changes in climate or in the levels of greenhouse gases, there is minimal spatial or taxonomic resolution. That is, all of the vegetation in a particular region is treated as a big leaf, with basically a uniform aggregate response "representative" of the plant community. But this can lead to serious problems of under-specification of detail.

Steve Pacala, Ben Bolker, Charles Canham, and I have been working with Fakhri Bazzaz, a plant physiologist at Harvard, who in his laboratory is studying the differential responses of plants to enhanced carbon. If one were to use his data to formulate an aggregate ("green slime") model and consider the effects of doubling CO_2, assuming fixed community composition, one would substantially underestimate the community response as compared to models that allow community composition to change in relation to plants' differential responses to fertilization. Thus, attention to the individualistic responses is essential. But how much?

For mobile animals, similar problems exist. One can study responses, say, of fish to an enhanced level of cadmium in laboratory situations to determine toxicity and other organismal responses. However, when one tries to move to field situations, one may find that fish simply swim away from polluted regions. The toxicity studies

1-56670-309-3/98/$0.00+$.50

by themselves are irrelevant. Thus we must find ways to extrapolate from the laboratory to the field, and to describe behavioral aspects, including individual movements.

To address such problems, one must take an individual-based approach. For example, data on the distribution of krill in the Antarctic show that, on very broad scales, the distribution of krill is well-explained by physical factors, the physics of the oceans, and correlates well with the fluorescence spectrum (Levin et al. 1989). However, on fine scales, the distributions are very different from one another: krill are much more patchily distributed because they orient toward one another, creating small-scale aggregations. Thus one needs an approach that takes into account factors on multiple scales, merging fluid dynamics on broad scales with individual dynamics on fine scales (Levin et al. 1989).

A hint regarding how to proceed comes from fluid dynamics. There are two ways to describe the behavior of fluids. One is the familiar Eulerian approach, in which one focuses on location in space and the local evolution of key descriptors. This is complemented by the Lagrangian approach, in which one essentially follows a packet of fluid as it makes its way around the system. Those two approaches are basically equivalent. Eileen Hofmann, at Old Dominion University, began to relate these for krill in the Antarctic, starting from the Eulerian description of fluid dynamics, the Navier–Stokes equations, and deriving representations of forces on individual particles. In this way, she could describe the movement of life stages that were being passively distributed, for example, the eggs of the krill. As krill progress from larval stages up to adult stages, they begin to swim and to orient toward each other; passive approaches no longer suffice. My student Daniel Grünbaum, in work at Cornell University (1992), dealt with this by adding inter-individual forces, ultimately returning to a full Eulerian description by an appropriate limiting process.

Global change presents a second example. Temperatures are rising; and depending on what scenario one adopts for introduction of materials into the environment, one may predict global mean temperature increases over the next 50 years. Obviously, that's very aggregated information. What can we say on more local levels? Predictions of this sort arise from gridded models, which break up the globe into cells of typically 10 degrees on a side, that is about 500 to 700 km on a side at temperate latitudes. More detailed models may be implemented, at great computational cost, and the capability to do so is increasing; but the resolution will not be better than one degree in the foreseeable future. Furthermore, even were we operating on a one degree grid, that would mean 50 to 70 km on a side, whereas most ecological information is collected at much finer scales. How do we go from ecological studies at the scale of tens of meters, or meters, or centimeters, up to the scales that can interact with climate change models? Models of forest response to climate change, or enhanced CO_2, typically begin from baseline models for the growth of plants, add shade and nutrient interactions, and then impose a mortality or growth filter based on laboratory or other data. Such models typically have limited spatial resolution, making difficult the prediction of effects on ecosystems.

Indeed a basic problem must be confronted at this juncture: What is an ecosystem? An ordinary toxicologist can focus on an organism, which gets sick or not, and dies or not. But an ecosystem has no such clear definition because individual species have individualistic distributions along gradients. Data collected at any particular temporal, spatial, or organizational scale provide only a slice of information, from an imposed perspective. We need to find how the dynamics depend on the scale of description, and how to relate phenomena on one scale to those on different scales. The most vexing problem is the hierarchical one: How should individuals be lumped into groups? Taxonomically? Functionally? With what resolution? There is no single correct answer, and we must explore how description varies with resolution, how responses and effects on ecosystems depend on the scales of description and aggregation, what the consequences are of changing those, and how phenomena at the level of individuals scale up to higher levels.

For forest response to climate change, we divide the landscape into a mosaic of cells. In each cell we have some dynamic responses of individual trees to multiple stressors and multiple influences, and then allow dispersal and movement among patches. Our aim is to get some idea of the overall response of the system to change and how its description depends on the scale of resolution. Our approach is to relate the pattern that we see in the distribution of species to individualistic responses. We seek to relate broad-scale information (remotely-sensed) to detailed information that has been collected at the level of individual plants, quantifying how they respond to stressors. Modeling serves as a bridge among levels. Our goal is to explain the patterns that we see on the broad scales in terms of responses on the fine scales, and to ask how much detail is needed at fine scales to explain broad-scale responses. How do we go from the individual to the ecosystem? How much information must we include, and how much can we throw away? The problem is one of aggregation and of simplification. The objective must be to develop appropriate and adequate macroscopic descriptors of system response and to investigate their dynamics.

In conclusion, the critical problem in ecotoxicology is one of scaling. How do effects on individuals become aggregated to populations and to species, and how much detail do we need to document species distributions? Until such issues can be addressed, the science of ecotoxicology will be unable to deal adequately with the preservation of our life-support systems.

REFERENCES

Grünbaum, D., *Local Processes and Global Patterns: Biomathematical Models of Bryozoan Feeding Currents and Density Dependent Aggregations in Antarctic Krill,* Dissertation, Cornell University, Ithaca, New York, 1992.

Levin, S. A., Morin, A., and Powell, T.H., Patterns and processes in the distribution and dynamics of Antarctic krill, in *Scientific Committee for the Conservation of Antarctic Marine Living Resources Selected Scientific Papers Part 1*, SC-CAMLR-SSP/5, CCAMLR, Hobart, Tasmania, Australia, 1989, pp. 281-299.

3 A Framework for Evaluating Organism Responses to Multiple Stressors: Mechanisms of Effect and Importance of Modifying Ecological Factors

S. Marshall Adams, K. D. Ham, and R. F. LeHew

INTRODUCTION

Aquatic ecosystems are complex entities that are controlled and regulated by a variety of physicochemical and biological processes (Figure 3.1). In addition, aquatic organisms experience a variety of natural and man-induced stressors, both of which vary spatially and temporally. High variability in environmental factors combined with synergistic and cumulative interactions of these factors in aquatic ecosystems complicate the interpretation and evaluation of the effects of contaminant-related stressors on organisms.

Responses of organisms to environmental conditions are the integrated result of direct and indirect contaminant impacts, natural environmental stressors (e.g., varying temperature and hydrologic regimes, sediment loading), or a combination of both natural and human-induced perturbations. Correlating specific contaminant-related responses at the lower levels of biological organization (i.e., biomolecular/biochemical) to effects at higher levels (e.g., growth, reproduction) is difficult because of the physicochemical and biological complexity of most aquatic systems. The extent to which laboratory tests alone are, or ever will be, capable of predicting either the likely exposure or the effects of chemical pollutants on ecosystems and their components has been questioned (Kimball and Levins 1985; Cairns 1986). Laboratory studies may provide limited information about simplistic links between a particular toxicant and specific organism responses, but most of these types of controlled studies cannot be reliably extrapolated to the natural environment because they lack ecological realism (NRCC 1985; Munkittrick and Dixon 1989). Test conditions of laboratory studies seldom accurately reflect the environment of natural populations (Cairns 1981), and water quality objectives have seldom been validated

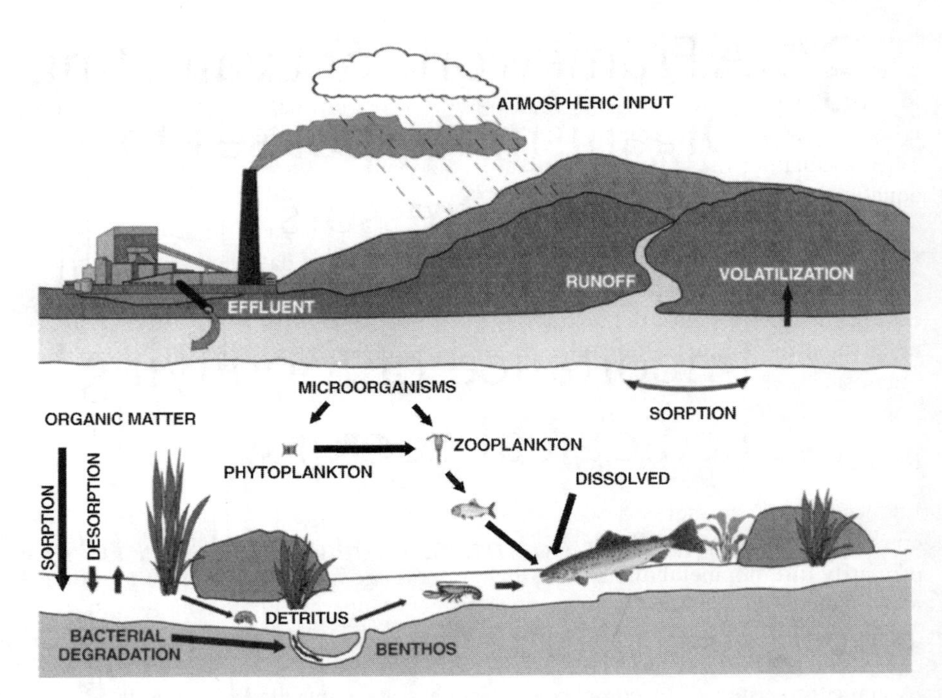

FIGURE 3.1 Simplified diagram of an aquatic ecosystem illustrating the main biological and physicochemical components of the system which have the potential to influence the response of fish to contaminant stressors.

in the field (Suter et al. 1985). These types of controlled or manipulative studies are generally not conducive to determining causal relationships between environmental stressors and responses at higher levels of biological organization because they do not take into account the many interacting and modulating factors in the natural environment (Depledge and Fossi 1994). Natural processes operating in the food web, such as interspecific and intraspecific competition, predator–prey relationships, and density-dependent interactions, could have major influences on the nature, magnitude, and final expression of a contaminant response in fish populations. In field situations, however, experimental conditions are difficult to control, and evaluation of causal relationships is generally circumstantial in nature, being based primarily on the weight-of-evidence approach (Suter et al. 1994). Also, in a natural field situation, the responses of organisms to environmental stressors, particularly at the higher levels of biological organization, are integrative in nature, reflecting the totality of the environmental conditions that impinge upon them (Ryder and Edwards 1985; Colby 1984).

Because of the complexity of natural systems, many approaches have proven to be inadequate for realistically assessing the effects of environmental stressors on aquatic ecosystem health. To help minimize some of the limitations of field studies in assessing cause and effect between stressors and biological responses, studies should also include factors that are useful for evaluating and interpreting the integrated health responses of these organisms. For example, nutrition and feeding, habitat influences, competition, and other noncontaminant-related parameters, such as temperature, varying hydrologic regimes, and turbidity, could all be important factors influencing the nature and magnitude of organismal responses to environmental stressors. Therefore, an attempt should be made to account for these factors when evaluating the relative contribution of contaminants to the health status of aquatic organisms.

The primary objective of this study is to provide an interpretative framework for evaluating the relative importance of direct and indirect pathways or mechanisms on the expression of pollutant responses in fish. In addition, the role of important ecological factors such as competition, food, and habitat availability in helping to understand the biological significance of contaminant-related stressors on aquatic organisms is also evaluated using examples from field data.

REGULATING MECHANISMS

Environmental stressors such as contaminants can affect fish populations and communities by two basic pathways or mechanisms. Direct mechanisms occur primarily through metabolic effects that are initiated at the lower levels of biological organization. Indirect mechanisms, however, operate mainly through effects on the food chain and on the behavior of organisms (Adams 1990). The effects of pollutants on organisms via direct mechanisms occur initially at the molecular or subcellular level and can be expressed, for example, as increases in mixed function oxidase (MFO) enzyme activity or DNA damage (Thomas 1990). Responses at this level can be propagated upward through increasing levels of biological complexity and affect physiological and metabolic processes such as lipid dynamics, immunocompetence, and hormone regulation (Larsson et al. 1985). Ultimately, these effects may be manifested as changes at the population and community level. In addition, pollutants may impact organisms indirectly through the food chain by influencing the quality (energy and protein content) and quantity (biomass) of energy available to consumers. Behavior, in turn, can be influenced by contaminants impairing reproductive, feeding, or habitat selection activities (Reynolds and Casterlin 1980). The more ecologically relevant parameters of aquatic systems, such as organism growth, reproduction, and population-level attributes, can therefore be affected by both direct and indirect mechanisms of contaminant stress and include the integrated effects of metabolic impairment, energy availability, and behavioral alterations (Adams 1990).

A helpful approach for achieving a better understanding of the relative importance of direct and indirect mechanisms in evaluating the responses of fish populations to multiple environmental stressors is to measure a selected suite of stress indicators representing three major categories of response indicators (Table 3.1). These three categories are (1) direct indicators of contaminant exposure, which include the MFO enzymes and DNA damage, (2) direct indicators of pollutant effects that primarily reflect metabolic dysfunction, such as serum enzymes and measures of lipid dynamics, and (3) indirect indicators of pollutant effects including nutrition and feeding indices, certain histopathologic indicators, and various measures of lipid pools within the organism.

TABLE 3.1
Major Categories of Contaminant Response Indicators and Some Representative Parameters Within Each Category that Could Be Used in Environmental Stress Studies to Help Evaluate the Relative Importance of Direct and Indirect Mechanisms in Influencing Stress Responses of Fish Populations

Direct Indicators of Toxicant Exposure	Direct Indicators of Metabolic Dysfunction	Indirect Indicators of Exposure
Detoxification enzymes	Serum enzymes (organ dysfunction)	Nutritional/Feeding
DNA damage	Lipid metabolism	Lipid pools
Body burdens of pollutants	Immunocompetence	Bioenergetic parameters
Selected histopathology — various necroses	Selected pathologies — melanophage aggregates	Bile color
Oxidative enzymes — peroxidases		Serum lipids (triglycerides and cholesterol)

A simplified example of how these various indicators could be used to evaluate the relative importance of direct vs. indirect mechanisms is illustrated in a situation where high levels of MFO enzymes (contaminant exposure indicators) and serum enzymes (organ dysfunction indicators) occur in an animal that is in good nutritional and bioenergetic condition. In this case, direct mechanisms of pollutant stress would be expected to have a relatively larger effect on organism response than would indirect mechanisms. Conversely, if indicators of pollutant exposure are low in organisms but their nutritional and bioenergetic status is impaired, indirect mechanisms may be implicated in influencing organismal response to stressors.

MODIFYING ECOLOGICAL FACTORS

A more realistic assessment of the relative importance of direct and indirect mechanisms in influencing the integrated responses of organisms to contaminant stress should also include an evaluation of the role of influential and modifying ecological factors such as food availability, competition (intraspecific and interspecific), and habitat quality (Adams et al. 1992). Knowledge of relative food availability is important because pollutants can dramatically affect both the quantity and quality of energy available to consumers (Munkittrick et al. 1991). Information about competition, as reflected in the relative abundance of competing species, and habitat availability (quality) is important because both of these ecological factors can affect how physiologically useful energy (assimilated energy) is partitioned among the main bioenergetic parameters of growth, reproduction, and energy storage. In situations where food availability is high, competition is low, and available habitat is good, increased metabolic costs associated with environmental stressors would suggest that direct mechanisms were primarily responsible for the observed stress responses. In con-

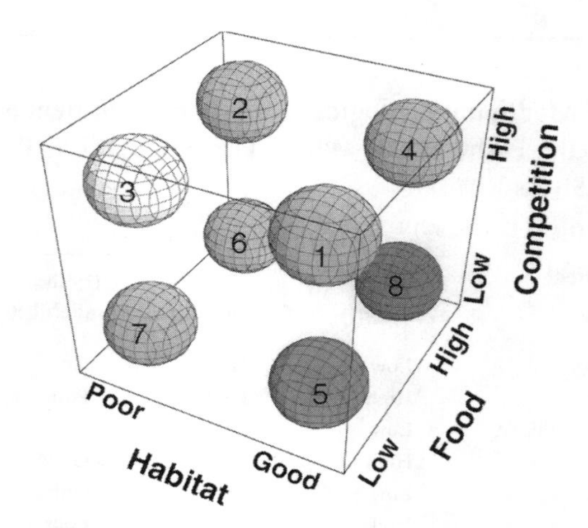

FIGURE 3.2 Principal ecological factors in an aquatic system which can influence or modify the responses of fish to contaminant stress. Combinations of ecological factors illustrated by the odd-numbered spheres suggest contaminant effects operating primarily through indirect pathways, and those factor combinations illustrated by the even-numbered spheres represent contaminant effects operating primarily through direct pathways. In general, the darker the sphere the higher the probability that contaminants themselves (either by direct or indirect pathways) have an effect on organism response.

trast, if stress responses in organisms occur under a situation in which food availability is low, competition is low, and habitat quality is good, then indirect pathways may be implicated as the primary mechanism involved in eliciting these responses.

In the aquatic environment, these ecological factors are present at different levels and combinations and can modify the expression of stress responses in fish. Some combinations of ecological factors that normally occur in field situations are illustrated in Figure 3.2. For example, in a situation of high food and good habitat availability and low competition (sphere 8, Figure 3.2), those stress effects in fish characterized by altered metabolic function and other biochemical or biomolecular changes would suggest that direct toxicant parameters were primarily responsible. However, under conditions of low food availability, reduced competition, and high habitat quality, indirect pathway effects would play an increasing role in influencing the observed stress responses (sphere 5, Figure 3.2). In this case, food availability is low even though competition for food is low, implying that some mechanism other than competition (such as contaminant effects on prey) may have reduced food availability. Contaminants are known to have major effects on the quality and quantity of food items for fish, such as benthic macroinvertebrates (Wallace et al. 1989; Wiederholm 1984). Alternatively, when food is low, competition is high, and habitat quality is poor, both direct and indirect mechanisms could influence the stress responses (sphere 3, Figure 3.2). Combinations of these modifying ecological factors that influence the primary pathways and mechanisms by which fish populations respond to contaminant stressors are provided in Table 3.2.

TABLE 3.2
Combinations of Modifying Ecological Factors that Influence the Level of Response and the Pathways by Which Fish Populations Respond to Contaminant Stress

Contaminant Effect		Modifying Ecological Factors			Corresponding sphere (in Figure 3.2)
Level	Pathway	Food availability	Competition	Habitat availability	
Moderate	Indirect	Low	High	Good	1
Moderate	Direct	High	High	Poor	2
Low	Direct & Indirect	Low	High	Poor	3
Moderate	Direct	High	High	Good	4
Primary	Indirect	Low	Low	Good	5
Moderate	Direct	High	Low	Poor	6
Moderate	Indirect	Low	Low	Poor	7
Primary	Direct	High	Low	Good	8

APPLICATION

To demonstrate how direct and indirect pathways (mechanisms) and modifying ecological factors can be used to help interpret stress responses in fish populations, we have applied these concepts to a stream ecosystem contaminated by industrial effluents. Individual and integrative measures of fish health were determined on redbreast sunfish (*Lepomis auritus*) collected from four sites (C1–C4) in a stream receiving point-source discharges of contaminants near its headwaters. Site C1 is immediately below the industrial outfall, C2 is 4 km downstream of C1, C3 is 9 km downstream of C1, and C4 is 17 km downstream of the outfall near the stream mouth. This system has a distinct gradient in contaminant loading, as evidenced by the decreasing downstream concentrations of PCBs and mercury in sunfish (Southworth 1990) and in sediments (TVA 1985). Fish at each of these four sites were analyzed for bioindicators of contaminant exposure (detoxification enzymes), organ dysfunction (blood chemistry), tissue and organ damage (histopathology), nutritional and bioenergetic responses, and overall health and condition (condition indices) (Adams and Ryon 1994).

The modifying ecological factors — food, habitat availability, and competition — were each evaluated as to their relative importance at each of the four sampling sites in the contaminated stream. As an estimate of competition (intraspecific and interspecific), the population density of each major fish species at each site was measured using a three pass-removal method (Ryon 1993; Carle and Strub 1978). Food availability was evaluated by measuring the density and biomass of benthic macroinvertebrates at each site (Smith and Tolbert 1993). Habitat quality at each site was estimated based on measurements of stream flow and substrate, bank cover, canopy (overhead vegetation), and the pool-to-riffle ratio (Ryon, et al. 1993).

TABLE 3.3
Primary Fish Stress Responses, Status of Ecological Factors, and Principal Stress Response Pathways Influenced by These Combinations of Factors for Fish Sampled from Four Sites in a Contaminated Stream

	Primary Fish Stress Response				Relative Availability		Competition	Principal Response Pathway
Site	Detoxification Enzymes	Lipid Metabolism	Liver Damage	Organ Dysfunction	Food	Habitat		
C1	P	P	P	A	Intermediate	Poor	High	Both direct and indirect
C2	P	P	P	P	Low	Fair	High	Mostly indirect via food chain; possible behavioral effects
C3	P	P	A	P	High	Fair	High	Mostly direct
C4	P	A	A	A	Intermediate	Good	Low	No physiological or histological stress responses observed

P = stress response present, A = stress response absent.

As evidenced by the induction of detoxification enzymes, fish at all four sampling sites were exposed to contaminants (Table 3.3). Expression of this exposure as stress responses in these fish was reflected primarily by altered lipid metabolism at the upper three sites (C1–C3), liver damage at C1 and C2, and organ dysfunction in fish sampled from the two intermediate sites (C2 and C3). At the lowest downstream site (C4), even though enzymes were elevated, no other stress responses were observed, indicating that either contaminants were not high enough to cause physiological impairment, or the MFO system of the fish was able to detoxify pollutants before any significant physiological or structural (i.e., histopathological) damage occurred. Notice, however, that liver damage was observed in fish from the upper two sites, possibly indicating that the MFO system might have been impaired somewhat in its ability to detoxify contaminants (Jimenez et al. 1990).

At the site nearest the industrial outfall (C1), biomass of available fish food (macroinvertebrates) was at intermediate levels, competition was high due to the relatively high abundance of several fish species at this site, and habitat quality was rather poor (Table 3.3). Based on this combination of ecological factors, it appears that direct effects operating through metabolic pathways and indirect mechanisms operating through the food chain may both be responsible for the responses observed in fish from this site (spheres 2 and 3, Figure 3.2 and Table 3.2). At the next downstream site (C2), competition was also high but food availability was low and habitat quality was fair. Under this combination of ecological factors, we might conclude that there is an increased probability that indirect effects were the principal mechanisms operating at this site (sphere 1, Figure 3.2 and Table 3.2). At site C3, competition was also high and habitat quality was fair but food availability was high. This particular combination of factors suggests that direct metabolic pathways may be the primary mechanism dictating stress responses of fish at this site (spheres 2 and 4, Figure 3.2) In this area of the stream, contaminants do not appear important in affecting food chain processes because even with high fish biomass (competition for food), food availability remains relatively high. At each of these sites a different pathway or mechanism appears to be involved in influencing the stress response of fish. Not only are different types or modes of stressors potentially involved at each site, but ecological factors also vary among sites which differentially modify these stress responses.

In conclusion, consideration of both direct and indirect mechanisms of pollutant stress along with a basic understanding of how principal ecological factors can modify organism response to contaminants should provide an interpretative framework for evaluating the effects of chronic stressors on aquatic organisms. Additional studies are necessary to better define the relative importance of ecological factors and how these factors interact with other environmental stressors in influencing contaminant-related stress responses in aquatic organisms.

ACKNOWLEDGMENT

The authors acknowledge the support of the Oak Ridge National Laboratory, managed by Martin Marietta Energy Systems, Inc., under contract DE-AC05-84OR21400 with the U.S. Department of Energy. Publication No. 4685 of the Environmental Sciences Division, Oak Ridge National Laboratory.

REFERENCES

Adams, S. M., Status and use of biological indicators for evaluating the effects of stress on fish, *Am. Fish Soc. Symp.*, 8, 1, 1990.

Adams, S. M. and Ryon, M. G., Comparison of health assessment approaches for evaluating the effects of contaminant-related stress on fish populations, *J. Aquat. Ecosyst. Health*, 3, 15, 1994.

Adams, S. M., Crumby, W. D., Greeley, M. S., Ryon, M. G., and Schilling, E. M., Relationships between physiological and fish population responses in a contaminated stream, *Environ. Toxicol. Chem.*, 11, 1549, 1992.

Cairns, J., Biological monitoring, Part IV — Future needs, *Water Res.*, 15, 941, 1981.

Cairns, J., The case for direct measurement of environmental responses to hazardous materials, *Water Res. Bull.*, 22, 841, 1986.

Carle, F. L. and Strub, M. R., A new method for estimating population size from removal data, *Biometrics*, 34, 621, 1978.

Colby, P. J., Appraising the status of fisheries: Rehabilitation techniques, in *Contaminant Effects on Fisheries*, Cairns, V. W., Hodson, P. V., and Nriagu, J. O., Eds., John Wiley & Sons, New York, 1984, chap. 16.

Depledge, M. H. and Fossi, M. C., The role of biomarkers in environmental assessment (2). Invertebrates, *Ecotoxicology*, 3, 161, 1994.

Jimenez, B. D., Oikarik, A., Adams, S. M., Hinton, D. E., and McCarthy, J. F., Hepatic enzymes as biomarkers: interpreting the effects of environmental, physiological, and toxicological variables, in *Biological Markers of Environmental Contamination*, Shugart, L. R. and McCarthy, J. F., Eds., Lewis Publishers, Chelsea, Michigan, 1990, chap. 6.

Kimball, K. D. and Levins, S. A., Limitations to laboratory bioassays: the need for ecosystem-level testing, *Bioscience*, 35, 165, 1985.

Larsson, A., Haux, C., and Sjobeck, M., Fish physiology and metal pollution: results and experiences from laboratory and field studies, *Ecotoxicol. Environ. Saf.*, 9, 250, 1985.

Munkittrick, K. R., and Dixon, D. G., Use of white sucker (*Catostomus commersoni*) populations to assess the health of aquatic ecosystems exposed to low-level contaminant stress, *Can. J. Fish. Aquat. Sci.*, 46, 1455, 1989.

Munkittrick, K. R., Miller, P. A., Barton, D. R., and Dixon, D. G., Altered performance of white sucker populations in the Manitouwadge chain of lakes is associated with changes in benthic macroinvertebrate communities as a result of copper and zinc contamination, *Ecotoxicol. Environ. Saf.*, 21, 318, 1991.

NRCC (National Research Council of Canada), The role of biochemical indicators in the assessment of ecosystem health — their development and validation, Publication 24371, National Research Council of Canada, Ontario, 1985.

Reynolds, W. W. and Casterlin, M. E., The role of behavior in biomonitoring of fishes: laboratory studies, in *Biological Monitoring of Fish*, Hocutt, C. H. and Stauffer, J. R., Eds., Lexington Books, Lexington, Massachusetts, 1980, chap. 4.

Ryder, R. A. and Edwards, C. J., A conceptual approach for the application of biological indicators of ecosystem quality in the Great Lakes Basin, Great Lakes Fisheries Commission International Joint Committee, Windsor, Ontario, 1985.

Ryon, M. G., Instream ecological monitoring — Fishes, in *Second Report on the Oak Ridge Y-12 Plant Biological Monitoring and Abatement Program for East Fork Poplar Creek*, Hinzman, R. L., Ed., Report Y/TS-888, Oak Ridge Y-12 Plant, Oak Ridge, Tennessee, 1993, chap. 6.2.

Ryon, M. G., Smith, J. G., and Peterson, M. J., Substrate and Cover, in *Second Report on the Oak Ridge Y-12 Plant Biological Monitoring and Abatement Program for East Fork Poplar Creek*, Hinzman, R. L., Ed., Report Y/TS-888, Oak Ridge Y-12 Plant, Oak Ridge, Tennessee, 1993, chap. 2.6.

Smith, J. G. and Tolbert, V. R., Instream ecological monitoring-Benthic macroinvertebrates, in *Second Report on the Oak Ridge Y-12 Plant Biological Monitoring and Abatement Program for East Fork Poplar Creek*, Hinzman, R. L., Ed., Report Y/TS-888, Oak Ridge Y-12 Plant, Oak Ridge, Tennessee, 1993, chap. 6.1.

Southworth, G. R., PCB concentration in stream sunfish (*Lepomis auritus* and *L. macrochirus*) in relation to proximity to chronic point sources, *Water Air Soil Pollut.*, 51, 287, 1990.

Suter, G. W., Barnthouse, L. W., Breck, J. E., Gardner, R. H., and O'Neill, R. V., Extrapolating from the laboratory to the field: how uncertain are you?, *Am. Soc. Test. Mater. Spec. Tech. Publ.*, 854, 400, 1985.

Suter, G. W., Sample, B. E., Jones, D. S., and Ashwood, T. L., *Approach and Strategy for Performing Ecological Risk Assessments for the U.S. Department of Energy's Oak Ridge Reservation,* Report ES/ER/TM-33/R1, Oak Ridge National Laboratory, Oak Ridge, Tennessee, 1994.

Thomas, P., Molecular and biochemical responses of fish to stressors and their potential use in environmental monitoring, *Am. Fish. Soc. Symp.*, 8, 9, 1990.

TVA (Tennessee Valley Authority), *Instream contaminant study — Sediment Characterization,* Vol. 1, Report Office Natural Resources Economic Development, Tennessee Valley Authority, Knoxville, Tennessee to the U.S. Department of Energy, Oak Ridge, Tennessee, 1985.

Wallace, J. B., Lugthart, G. J., Cuffney, T. F., and Schurr, G. A., The impact of repeated insecticidal treatments on drift and benthos of a headwater stream, *Hydrobiologia*, 179, 135, 1989.

Wiederholm, T., Responses of aquatic insects to environmental pollution, in *The Ecology of Aquatic Insects*, Resh, V. H. and Rosenberg, D. M., Eds., Praeger Publishers, New York, 1984, chap. 17.

4 Forest Ecosystems and Air Pollution: The Importance of Multiple Stress Interactions on a Regional and Global Scale

George E. Taylor, Jr.

INTRODUCTION

In the last several decades, it has become evident that anthropogenic activities are having a profound influence on the earth's surface. One of the more demonstrable effects has been changes in the concentrations of trace chemicals in the atmosphere on a local, regional, and global scale. Some of the more important examples are sulfur oxides (e.g., sulfate, SO_4; sulfur dioxide, SO_2), nitrogen oxides (nitrogen dioxide, NO_2; nitric oxide, NO; and nitric acid vapor, HNO_3), methane (CH_4), carbon monoxide (CO), particulates, hydroxyl radical (OH), volatile organic hydrocarbons (VOCs), carbon dioxide (CO_2), mercury (Hg), lead (Pb), tropospheric ozone (O_3), and stratospheric O_3. From the perspective of the environmental sciences, these changes are important because singly or in combination they can affect the (1) chemistry of the atmosphere, (2) physics of the atmosphere (i.e., global warming; visibility in pristine areas), (3) biology of at-risk organisms in either human or nonhuman species, and (4) structure/function of managed and unmanaged ecosystems (e.g., geochemistry of watersheds or surface waters).

This chapter focuses on those atmospheric pollutants and contaminants which are deposited to forest canopies and (1) affect tree physiology or plant community dynamics, (2) influence biogeochemistry at the stand and watershed level, or (3) are cycled through or altered by forests and subsequently influence human health (e.g., food chain transport). The emphasis on forest ecosystems reflects the evolution over the last decade in air pollution studies away from short-term (single growing season) air pollution impacts in monospecific stands of highly domesticated annual species toward chronic-level effects in unmanaged ecosystems (e.g., Tingey et al. 1990; Shaver et al. 1994). This evolution has required a different suite of analyses and expertise, reflecting the greater complexity in how natural ecosystems respond to stress relative to the response of agricultural systems. One of the more salient findings is that air pollution is only one of several stresses affecting forests and that most at-risk trees, stands, and ecosystems are challenged by a mix of natural and anthropogenic stresses. Because the mix of stresses varies spatially and temporally in unmanaged landscapes far more than in agricultural ecosystems, the investigations by ecologists are very different from those conducted by plant pathologists in agricultural settings (Taylor et al. 1994).

The objectives of this chapter are:

- to describe the attributes of air pollution that are important to investigating ecological effects in forests
- to describe some of the methodologies used to address air pollution effects in forests
- to discuss the role of modeling as a tool to address air pollution effects at multiple temporal and spatial scales
- to provide a case study demonstrating the presence of multiple air pollutants in many North American forests

There are a number of recent monographs that explore in considerable detail various aspects of the interactions of air pollution and forests, and these include Schulze et al. (1989), Eagar and Adams (1992), Johnson and Lindberg (1992), Smith (1990), Olson et al. (1992), Barker and Tingey (1992), and Last and Watling (1991).

ECOLOGICAL ATTRIBUTES OF AIR POLLUTION IN FORESTS

The interaction of air pollution and forests is really an ecological and plant physiological issue. Unfortunately, the science of air pollution effects emerged three to four decades ago within the discipline of plant pathology, and this origin has significantly influenced the development of the science in such areas as exposure methodologies, injury assessment, statistical analyses, and application of models. Only within the last decade have more ecological and physiological underpinnings begun to displace the perspective of plant pathology as a basis for assessing effects (e.g., Pell et al. 1994; Laurence et al. 1994). Thus, it is important to reiterate the aspects of air pollution and forest responses that are based on ecological principles, and let these serve as the basis for future investigations. The following is a synopsis of some of the more prominent ecological underpinnings.

A fundamental feature of studies investigating air pollution and forest ecosystems is the importance of atmosphere–biosphere interactions (Hosker and Lindberg 1982; Johnson and Lindberg 1992). This dictates an emphasis on atmospheric and meteorological processes not commonly addressed in ecology, and the most important are those processes governing pollutant fluxes from the atmosphere to the biosphere (Lovett 1994). This includes deposition of airborne chemicals to terrestrial landscapes via wet (rainfall and cloud water) and dry (particle and gas) deposition. In addition, the concept of pollutant exposure dynamics (temporal and spatial patterns of concentrations) is critical in determining effects on the physiology, growth,

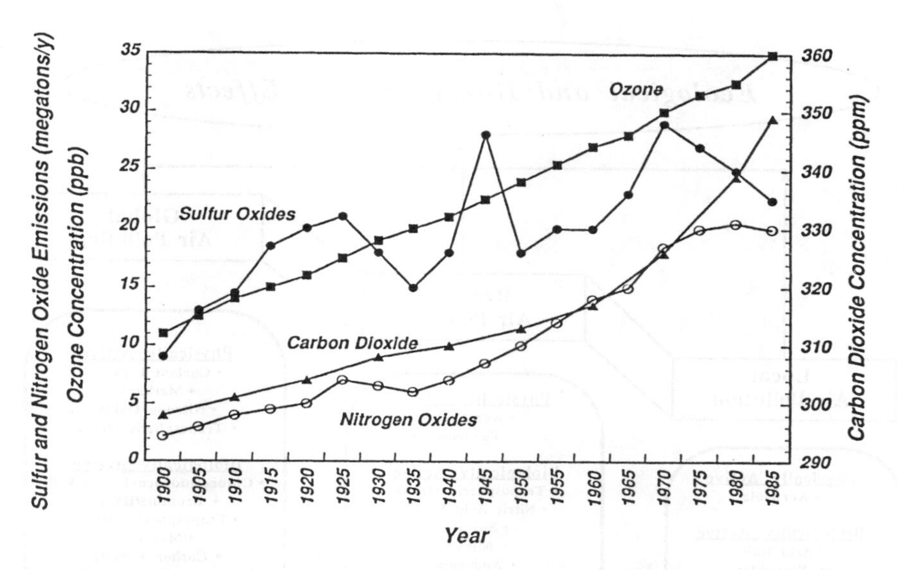

FIGURE 4.1 Changes in the chemistry of the atmosphere over the last century in North America. The data for CO_2 are from Keeling et al. (1989), O_3 from Taylor et al. (1994), and emissions of oxides of nitrogen and sulfur from McLaughlin and Norby (1991).

and reproductive success of sensitive species (Hogsett et al. 1988; Taylor and Hanson 1992).

Until recently, concerns over air pollution focused largely on individual chemicals causing visible or dramatic effects on community dynamics and ecosystem structure and/or function (Gordon and Gorham 1963; Smith 1990). This emphasis has been replaced by a recognition that most terrestrial ecosystems are not impacted solely by one pollutant, that exposure and deposition are chronic rather than acute, and that effects in natural stands are largely the result of the interaction of air pollution with the suite of other environmental stresses (McLaughlin 1985; Mooney et al. 1991).

For example, it is widely recognized that the ~ 10% increase in CO_2 over the last century from 295 to 350 ppm (Figure 4.1) has affected terrestrial vegetation in general and forests in particular (e.g., Bazzaz 1990). Simultaneous with this 10% increase in CO_2 on a global scale, there has been a 350% increase in tropospheric O_3 from approximately 10 ppbv to 35 ppbv in North America (Figure 4.1). Ozone is well documented to affect the physiology of at-risk forest tree species on a regional, hemispherical, and global scale (Taylor et al. 1994). Concurrent with these increases in CO_2 and tropospheric O_3 in North America in the last century, there has also been a pronounced increase in emissions of sulfur and nitrogen oxides by a factor of 2 and 10, respectively (Figure 4.1). Forests are responding to many chemical changes in the atmosphere in addition to the array of other factors (e.g., drought, land use) that also affect forest trees, stands, and watersheds. It is inappropriate to attribute changes in forest tree physiology or stand ecology within a given area to one given stress, either of anthropogenic or natural origin, without careful accounting of the other environmental stresses.

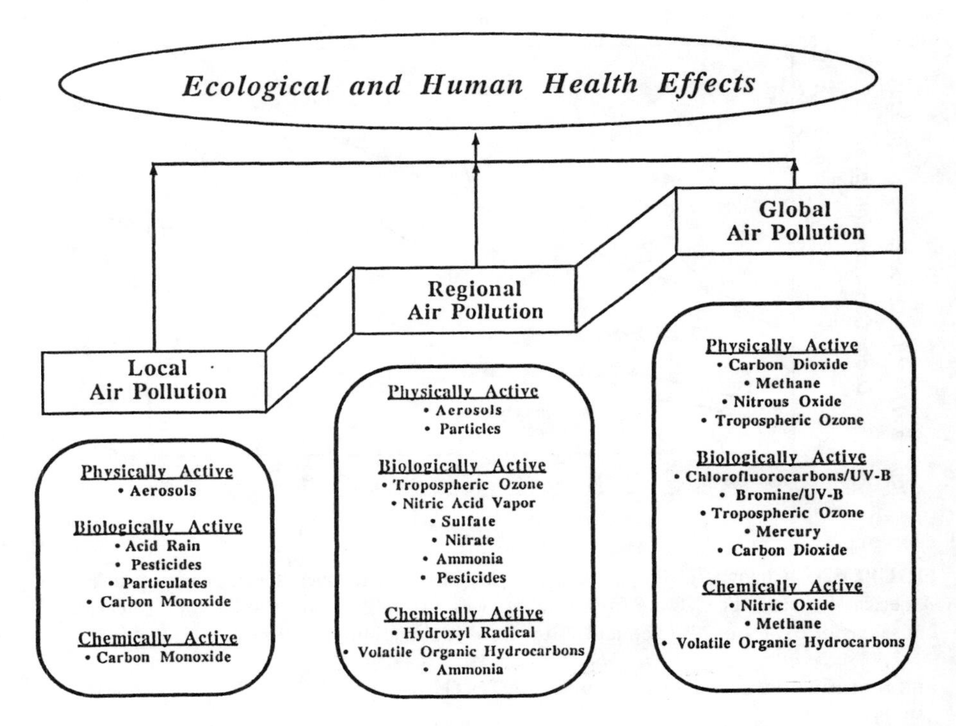

FIGURE 4.2 Local, regional, and global distribution of air pollution. Each panel shows the principal trace gases, aerosols, and particulates at the three scales and whether the effects are due to physical (e.g., radiatively active; visibility), biological (effects on human health, agriculture, or ecological systems), and/or chemical processes (i.e., reactions in the atmosphere). Irrespective of the process affected by the pollutant, the ultimate concern is effects on ecological systems (e.g., forests) and human health.

Another important attribute of air pollution is the spatial and temporal distribution. Whereas some forests are impacted by local emissions, deposition of most pollutants to forests occurs after long-range atmospheric transport; thus the distribution of the stress ranges from one of local to global in scale (Figure 4.2). Thus, landscapes across which deposition occurs vary depending on climate, geology, land use, and human intervention, and identifying specific source-receptor relationships across these broad and fragmented landscapes has a low probability of success. Retrospective analyses have shown that changes in many chemicals occur over time frames of several decades to a century in duration (Table 4.1) and that annual increments of change are commonly <1% (e.g., Farman et al. 1985; Volz and Kley 1988; Swain et al. 1992; Thompson 1992). Thus, the emphasis is both ecological and physiological consequences of small cumulative changes in pollution over time horizons that approach several to many decades. One aspect of the retrospective analyses is particularly important: most airborne pollutants clearly existed in the pre-industrial atmosphere due solely to natural processes (Table 4.1). The consequence is that biota in terrestrial landscapes are not challenged by novel chemical inputs.

TABLE 4.1
Some Atmospheric Pollutants of Interest to Ecologists

Category	Pollutant	Mean Residence Time	Distribution	Source[a]
Carbon	CH_4	Years	Globe	A&N
	VOCs	Days	Region	A&N
	CO_2	Decades	Globe	A&N
	CO	Days	Hemisphere	A&N
Nitrogen	NO	Days	Region	A&N
	NO_2	Days	Region	A&N
	HNO_3	Hours/Days	Region	A&N
	NH_3	Hours/Days	Region	A&N
	N_2O	Years	Globe	A&N
Sulfur	SO_2	Days	Region	A&N
Chlorofluorocarbons	CH_3CCl_3	Century	Globe	A
(CFCs)	CF_2Cl_2	Century	Globe	A
Halons	$CBrF_3$	Decades	Globe	A
Miscellaneous	O_3	Days	Region	A&N
	H_2O	Days	Region	N
	Hg^0	Years	Globe	A&N
	CH_3Br	Decades	Globe	A

[a] N = natural; A = anthropogenic

A variety of anthropogenic changes in pollutant concentrations are recognized by atmospheric scientists. However, only a subset is ecologically important. Within this subset, some compounds elicit toxic or harmful effects via inhibition of physiological function (e.g., tropospheric O_3). Depending on the deposition rate and nutrient status of the ecosystem, another class may enhance physiological function ("beneficial"), providing an element that naturally exists below optimum levels for growth and reproduction. The most notable example of this set of compounds is nitrogen, which is deposited to forests in multiple forms of wet and dry deposition. Kauppi et al. (1992) have argued that nitrogen deposition at a regional level is responsible for increases in stand productivity in Europe and possibly elsewhere. These pollutants are an environmental stress in an ecological context because they represent an unnatural change (Fenn and Bytnerowicz 1993). Fox et al. (1989) and Shaver et al. (1994) propose that in some landscapes (e.g., wilderness areas) any change in pollution levels is a stress, and this proposal is also the basis for the concept of critical pollution levels and incipient effects on sensitive plants (Cape 1993; Bull 1991). Last, some pollutants are of concern to ecologists because their deposition and subsequent behavior in continental landscapes result in xenobiotics being bioconcentrated and/or transferred in the food chain (Schroeder and Lane 1988; Zillioux et al. 1993; Fitzgerald and Clarkson 1991).

Direct effects of some air pollutants on terrestrial vegetation are well documented in intensively managed ecosystems and are commonly manifested as stress-specific symptoms of injury (e.g., foliar chlorosis). However, indirect effects are more important for vegetation in unmanaged natural ecosystems (Johnson and Taylor 1989), and these effects are defined as ones in which the pollutant alters the physical or chemical environment (e.g., soil properties) or the plant's ability to compete for limited resources (e.g., nutrients, water, light) or to withstand pests or pathogens (e.g., Fowler et al. 1989; DeHays 1992; Temple et al. 1993). The significance to ecologists is that air pollution can affect a species through both direct and indirect modes of action, and indirect effects are often of greater significance in unmanaged ecosystems.

The final feature of air pollution is the chemical's residence time in an ecosystem and processes that dictate whether the pollutant is conserved or ephemeral. Conserved species are ones in which the chemical or its reaction product moves through landscapes and is not transformed or mineralized; examples include atmospheric inputs of hydrogen ion, sulfur, nitrogen, Pb, and Hg (Lindberg et al. 1992). Incremental additions of these pollutants may result in progressive increases in concentration or depletion of related nutrients. The most important class of chemicals includes those that bioconcentrate and bioaccumulate (Travis and Hester 1991). For the ephemeral species, such as tropospheric O_3 (Taylor and Hanson 1992), the residence time of the chemical in the ecosystem is typically measured in time frames of minutes or less; the chemical's ecological significance lies more in direct and indirect toxicity to biota rather than in issues of biogeochemistry and food chain transfer.

EXPERIMENTAL METHODOLOGIES TO INVESTIGATE AIR POLLUTION EFFECTS ON FORESTS

Until recently, air pollution research on terrestrial vegetation was largely driven in a methodological sense by concerns over pollution impacts on agricultural crops (e.g., Heck et al. 1983; Heck et al. 1988). The methodology developed in response to the need for an economic endpoint, underpinned by single-factor exposure–response relationships. The ecological and physiological context of how natural ecosystems, especially forests, respond to air pollution required different methods. The reasons for this are many, including recognition of multiple pollutant deposition, long generation time of forest trees, genetic heterogeneity of the forests, multiple age classes and species, mixture of dissimilar landscapes, importance of indirect and cumulative effects, and concerns for food chain transport. Over the last decade, this recognition has prompted the development and application of new methodologies which take into account the uniqueness of unmanaged forest ecosystems, their responsiveness to air pollution, and the role of multiple stress interactions.

The development of methodologies for exposing trees to pollutants in both field and laboratory environments has been reviewed (Hogsett et al. 1987a and b). The principal performance criterion was control of pollutant exposure in a precise and reproducible manner. More recently, interest in air pollution concerns in the ecological sciences has prompted the development of even newer methods to maintain an environment that is near ambient in edaphic, climatic, and atmospheric attributes in addition to controlling the air pollutant exposure.

A variety of methods for air pollution studies in highly controlled laboratory conditions exist for both wet and dry deposited pollutants. Cuvettes at the individual leaf or branch are used on a limited basis for studies of pollutant flux, effects on stomatal conductance and photosynthesis (Hallgren et al. 1982; Ennis et al. 1990), and uptake of stable and radioisotopes to monitor the transport and fate of wet or dry deposited sulfur or nitrogen (Hanson and Garten 1992). For studies at the seedling level, chamber facilities exist for use in glasshouse environments or artificially lit growth rooms; the only unusual feature is instrumentation to control pollutant exposure. Continuously stirred tank reactors represent the most highly controlled of the indoor chambers (Rogers et al. 1977). Laboratory methodologies to simulate and/or exclude pollutants deposited via rainfall and cloud-water impaction are less common, and the uniqueness of these systems lies in the methods for droplet generation and dispensing (e.g., Skeffington and Roberts 1985). None of the systems is capable of simulating realistically the exposure dynamics of wet deposition, particularly in high elevation forests, where cloud-water impaction is the principal vector for pollutant deposition.

One of the most widely used field methods is the open-top chamber (Heagle et al. 1973), which provides near-ambient conditions of light intensity, photoperiod, and temperature while allowing control of the gaseous composition of air surrounding the canopy. Many chambers have covers to exclude ambient rainfall and are fitted with nozzles to simulate deposition of pollutants in rain (Johnston et al. 1986). The chambers have several advantages, including the ability to expose large numbers of plants simultaneously, vary peak concentration and duration to evaluate exposure–response relationships, manipulate several pollutant variables simultaneously, and achieve below-ambient exposures using filtration.

The environment within open-top chambers differs from that of ambient conditions, principally with respect to temperature, radiation, and atmospheric turbulence (Hogsett et al. 1987a and b). Consequently, trees grown inside chambers differ from their counterparts in nonchambered plots due solely to the inadvertent change in the physical environment (Olszyk et al. 1992).

Smaller chambers have been used to expose individual branches to pollutants under field conditions (Teskey et al. 1991), and these systems have several advantages. By exposing branches within the canopy of a single tree to different air pollution exposures, genetic variation in response is eliminated, and the photosynthetic and carbohydrate allocation responses to pollutant exposure can be compared. The major criticisms of the branch chamber methods are the degree to which branches within a canopy are autonomous for water, nutrients, and carbon and the importance of shoot–root feedback processes.

Nonchambered systems allow trees or part of the plant canopy to be exposed to at least some natural environmental conditions while also varying concentrations of gaseous pollutants, either by perfusing the canopy with filtered air or by adding the pollutant to the atmosphere and allowing atmospheric transport and turbulence to dilute and distribute the pollutant within the canopy (Hogsett et al. 1987a). Early exposure systems were characterized by high variability in exposure concentrations both vertically and horizontally, although a number of recent design modifications have rectified many of the problems (Tjoelker et al. 1994).

The availability of programmable control systems has led to the development of methodologies at a small stand level to vary pollutant injection frequency and concentration based on prevailing wind speed and direction (e.g., McLeod et al. 1985; Hendry et al. 1992). The primary advantages of free-air release systems are the elimination of artifacts associated with chambered systems and the allowance for more natural environmental factors (e.g., pests, pathogens, soil properties, climate conditions) to interact during the exposure. These systems have disadvantages, including a high degree of spatial and temporal variability in pollutant concentrations, inability to replicate the environment during exposure with respect to light, temperature, and humidity, and limitation of exposures to ambient and above-ambient treatments only.

There are a number of methodologies for assessing air pollution effects on forests under natural stand conditions. Air pollution, like many other stresses, causes biochemical and physiological changes in above-ground or below-ground organs. Rapid identification of stress-specific changes provides an opportunity not only to understand mechanisms of injury but also to identify stands at risk and the pollutants eliciting the injury (National Academy of Sciences 1989). Biomarkers are specific changes in cellular and/or biochemical processes that demonstrate a measurable and potentially significant change in the physiology and growth of an individual tree and that can be attributed to specific pollutants (e.g., Jones and Coleman 1989; Friend et al. 1992; Tandy et al. 1989).

Another field technique is comparative stand analysis (Cole et al. 1991), which is comparable to that of epidemiology in human health studies, and attempts to establish a correlation between indicators of forest stand structure or function and pollutant exposure. The dependent variable is a quantifiable gradient in pollutant exposure coupled with a minimum of concurrent gradients in other environmental parameters (Westman 1979; Miller 1973). The limitation of this methodology is the absence of ecologically meaningful indicators of forest health that can lead to a clear cause-and-effect relationship.

Radial increment growth is also sensitive to anthropogenic stresses including air pollution (McLaughlin et al. 1987; Peterson et al. 1993). Regardless of whether radial increment analysis isolates a single causative agent, the approach is useful for identifying shifts in stand-level productivity due to air pollution over time scales approaching decades to centuries. A related field technique is remote sensing, which can be a powerful tool for studying vegetation response to air pollution and for scaling from individual tree to forest stands to ecosystems (Voglemann 1988). Community canopy characteristics can be evaluated for unique spectral characteristics using various wavelengths in the visible and near-infrared regions.

Nutrient cycling in forest ecosystems has developed its own suite of experimental methodologies to understand the relationship between nutrient cycles and productivity. Air pollution researchers concerned with soil and water acidification realized that a biogeochemical cycling approach was necessary in order to account for the various sources and sinks of hydrogen ion (Ulrich 1980; Johnson and Lindberg 1992). The most notable change in field methodology with respect to air pollution studies has been the demonstration that atmospheric inputs of hydrogen ion, sulfur, nitrogen, and some base cations can be significant and that the forms of input can vary significantly from site to site, due to orographic factors and proximity to pollutant sources. New methodologies for more precise collection of atmospheric deposition have substantially altered estimates of nutrient fluxes in many forest ecosystems, and this is largely due to the degree to which measures of bulk precipitation inputs underestimate total deposition via wet and dry processes (Lindberg et al. 1986).

ROLE OF MODELING IN INVESTIGATING ECOLOGICAL EFFECTS IN FORESTS

Pollution effects begin with changes in tree physiology and growth and are manifested at the community level by alteration of competitive relationships and at the ecosystem level by modifications of nutrient cycling and productivity. The large number of variables that can be directly or indirectly affected by air pollution constrains the assessment of pollution effects by experimentation alone. An alternative is to use models of forest trees and stands to predict how forest resources will respond to existing or projected pollution scenarios.

Modeling has become an essential tool in the study of air pollution effects on forests (Kiester 1990; Dixon et al. 1990; Weinstein and Yanai 1994). Empirical or descriptive models have been used for decades to predict growth of a species under a defined range of time and site conditions. Process-based models are useful in the study of forest responses to stress since they are based on mechanisms and incorporate processes by which stresses of both natural and anthropogenic origins interact.

Most of the physiologically based process-models are modular and have detailed code for gross and net photosynthesis to simulate the effect of stresses that directly influence carbon acquisition. The subsequent allocation of photosynthate and partitioning of dry matter are less mechanistic, and most use general source–sink relationships and species-specific phenology to address carbon and nutrient allocation throughout the plant. Most models either ignore or treat inadequately below-ground processes, due to our relatively poor understanding of the physiology and biochemistry of roots and the effects of anthropogenic factors on the rhizosphere. At the individual tree level, examples of mechanistic models used in air pollution studies are TREGRO (Weinstein et al. 1991; Weinstein and Yanai 1994) and MAESTRO (Norman and Welles 1983). One of the more common shortcomings of some models is a preoccupation with the response of net photosynthesis, when in reality many data sets indicate that photosynthesis is unchanged in chronic air pollution situations or may even increase as a compensatory response (Pell et al. 1994).

The dynamics of natural stand succession have been investigated using gap-succession models (Botkin et al. 1972; Shugart et al. 1992), which simulate growth, reproduction, and mortality of individual trees in multispecies patches and predict the effects of competition. These models share the same basic characteristic equations for calculating growth, influences of limiting resource availability, establishment, and mortality in a forest stand. Examples of stand level models used to investigate air pollution effects are FORET (West et al. 1980) and ZELIG (Urban et al. 1991).

One of the most promising methodologies for investigating air pollution effects on regional forest resources uses models at different hierarchical scales. The model output based on the response of individual trees (time scale of years) serves as the input to a stand-level simulator. Thereafter, the output from the stand level (time scale of decades) serves as input into a geographic information system (GIS), allowing the matrix of interactive effects to be simulated at the landscape level, depending on the suite of natural and anthropogenic factors.

Nutrient cycling in forests has been modeled for more than three decades. Forest nutrient cycling models which investigate the influence of air pollution are not as well developed as those based on the growth of individual trees and forest stands. This shortcoming is changing as the community of forest scientists has recognized that atmospheric inputs of conserved chemicals can be quantitatively significant and that annual inputs over time can affect the biogeochemistry of forest stands (Schulze et al. 1989) or contribute to food chain transport and contamination. An example is the nutrient cycling model NuCM (Liu et al. 1991), which is one of the more recently developed simulators for studying the effects of air pollution on forest biogeochemistry. The code simulates forest canopy deposition of pollutants (via gas diffusion, particle sedimentation, rainfall, and cloud-water impaction), vegetation growth, litterfall and decay, soil biogeochemistry, and soil moisture hydraulics.

CASE STUDY OF MULTIPLE AIR POLLUTION EFFECTS IN FORESTS OF NORTH AMERICA

Increasing recognition is being given in environmental sciences to the role of multiple stresses in governing the productivity of vegetation in terrestrial landscapes, particularly in issues for which anthropogenic stresses are thought to be important. Recently, the U.S. program on acidic deposition concluded that some forest ecosystems in North America are intrinsically sensitive to either acidic deposition or O_3 (Barnard and Lucier 1990). Parallel studies of tropospheric chemistry concluded that hydrogen ion deposition and multiple indices of O_3 air quality in North America were not correlated spatially or temporally such that the two stresses do not co-occur in forests. This conclusion must recognize that there is a limited number of forested sites having co-monitoring instrumentation and situated in regions of ecological concern. A further complication is the reliance on rainfall as the sole vector for atmospheric deposition of pollutants. In many terrestrial landscapes, total deposition of hydrogen ion, nitrogen, and sulfur is largely determined by dry deposition of pollutant gases (e.g., nitric acid vapor) in low elevations and cloud-water interception in high elevations (Johnson and Lindberg 1992; Lovett 1994).

The co-occurrence of tropospheric O_3 and atmospheric deposition of hydrogen ion, sulfur, and nitrogen can be investigated using the data generated in the Integrated

Forest Study (Johnson and Lindberg 1992). This study addressed the role of atmospheric deposition on nutrient cycling processes in 10 forested ecosystems in North America that differed in a range of characteristics including proximity to pollutant sources, latitude (29 to 47 °N), elevation (100 to 1250 m), forest cover type (conifer to mixed hardwood), leaf surface area (4 to 11 m^2 m^{-2}), and synoptic-scale meteorological conditions that determine airshed dynamics. The comparative approach was adopted to help elucidate anthropogenic effects on patterns and rates of atmospheric deposition and nutrient cycling in forested ecosystems at different levels of risks from air pollution stress.

Each of the sites was instrumented with an array of meteorological sensors, wet and dryfall collectors (above canopy), filter pack samplers for trace gases and particles, and an O_3 analyzer (Johnson and Lindberg 1992). Annual atmospheric deposition to each of the forest canopies for 1987 and 1988 was calculated as the total derived from (1) precipitation (event-only collectors), (2) cloud-water interception of radiation fogs or orographically driven liquid water, and (3) dry deposition of gases (i.e., HNO_3, SO_2), and particles/aerosols (e.g., SO_4) estimated by the inferential method (Lovett 1994). In this methodology, atmospheric concentrations of trace gases and particles (by size) collected solely during dry periods were multiplied by a deposition velocity calculated on a site-specific basis that accounts for atmospheric turbulence, the canopy's physiological state (stomatal resistance), leaf surface wetness, and an array of climate conditions. Ozone concentrations were monitored continuously at each site according to U.S. EPA protocol and archived daily as twenty-four 1-hour means during the nominal April–September growing season (Johnson and Lindberg 1992). The statistic for O_3 analysis (SUM06) focused solely on the relationship between O_3 concentration and biological effects. This statistic addresses only those periods in which the O_3 concentration equals or exceeds a threshold value of 60 ppbv and is based on studies that suggest a disproportionate role for episodically high O_3 concentrations in governing the physiology and growth of terrestrial vegetation (Hogsett et al. 1988).

The results support the hypothesis that O_3 concentrations in forested landscapes of North America are positively correlated with atmospheric deposition of hydrogen ion, sulfate, and nitrate. With respect to solely hydrogen ion, the pairwise correlation coefficient was strongly positive ($r = +0.83$) (Figure 4.3). A similar pattern was evident for sulfate deposition; the correlation was again positive and very high ($r = +0.82$) (Figure 4.3). The relationship between nitrate deposition to forest canopies and O_3 exposure was consistent with that for either hydrogen ion or sulfate: the correlation was $r = +0.80$.

These data originated from the best array of atmospheric monitoring sites in forested locations in North America and addressed the issue of atmospheric deposition of multiple pollutants. The data clearly demonstrate that these forest sites experience a pattern in which inputs of nitrogen, sulfur, and hydrogen ion are correlated with one another and that these chemical inputs are in turn associated with exposure to tropospheric O_3. The O_3 exposures in North America are documented to be sufficient in many regions to influence the physiology and growth of sensitive forest species (Taylor et al. 1994) as the SUM06 exposure exceeds

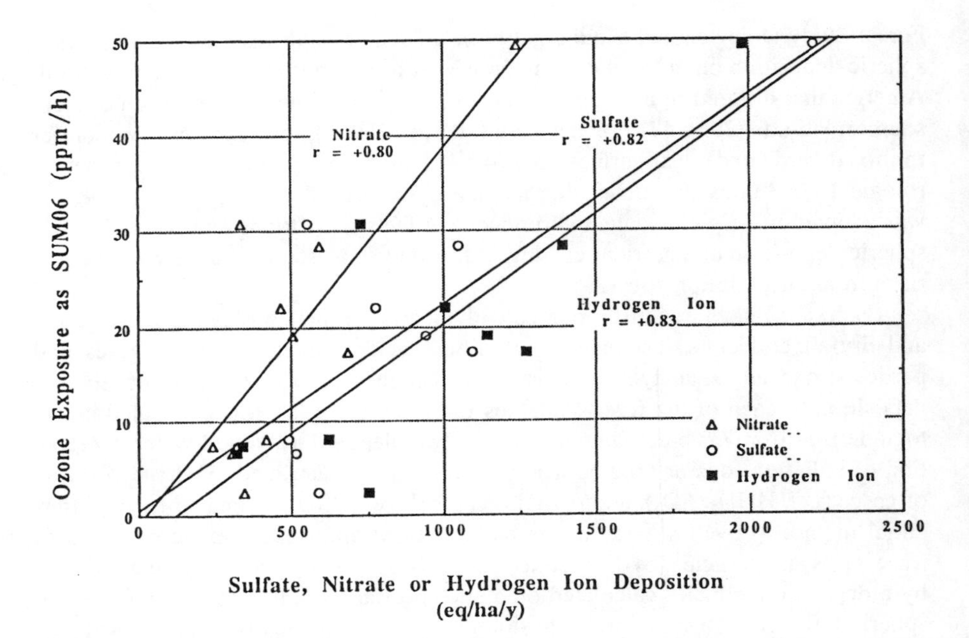

FIGURE 4.3 Relationship between ozone exposure in the forest of the Integrated Forest Study (Johnson and Lindberg, 1992) in North America and the deposition of sulfate, nitrate, and hydrogen ion. The deposition data represent the sum of all wet and dry deposition processes, including inputs of gases, particles, aerosols, rain, and cloud water. The relationship between the paired variables is presented as a linear correlation analysis and demonstrates that in all cases the correlations (r) are positive and equal to or greater than 0.08. This indicates that the input of chemicals to forests of North America tends to be correlated so that forests at risk from ozone exposure are also experiencing high inputs of nitrogen, sulfur, and hydrogen ion.

25 ppm/h. The additional inputs of nitrate, sulfate, and hydrogen ion are of concern ecologically in many landscapes (e.g., Bull 1991; Cape 1993) although the degree to which these inputs constitute a long-term stress is site dependent (Johnson and Lindberg 1992; Eagar and Adams 1992). The salient point is that most forests are being impacted by several atmospheric pollutants in a subtle, chronic manner and that investigations of the ecology of these ecosystems must recognize how these stresses of anthropogenic origin interact with the suite of more natural stresses.

ACKNOWLEDGMENTS

The author acknowledges with appreciation support from the U.S. Environmental Protection Agency's Cooperative Agreement entitled, "Ozone Forest Response Program," the National Institute of Environmental and Health Sciences grant entitled "Chemical Environmental Problems Associated with Historical and Current Precious Metal Mining," and USDA Project No. NEV00522M entitled, "Atmospheric-Forest Exchange of Nitrogen-Containing Trace Cases in the Sierra Nevada Landscape."

REFERENCES

Barker, J. R. and Tingey, D. T., Eds., *Air Pollution Effects on Biodiversity,* Van Nostrand Reinhold, New York, 1992.

Barnard, J. E. and Lucier, A. A., Changes in forest health and productivity in the United States and Canada, in *Acidic Deposition: State of Science and Technology, Volume III: Terrestrial, Materials, Health and Visibility Effects*, Irving, P. M., Ed., U.S. National Acid Precipitation Assessment Program, Washington, D.C., 1990, pp. 25-185.

Bazzaz, F. A., The response of natural ecosystems to the rising global CO_2 levels, *Annual Review Ecology Systematics*, 21, 167, 1990.

Botkin, D. B., Frank, J. F., and Walls, J. R., Some ecological consequences of a computer model of forest growth, *J. Ecol.*, 60, 849, 1972.

Bull, K. R., The critical loads/levels approach to gaseous pollutant emission controls, *Environ. Poll.*, 69, 105, 1991.

Cape, J. N., Direct damage to vegetation caused by acid rain and polluted cloud: definition of critical levels for forest trees, *Environ. Poll.*, 82, 167, 1993.

Cole, J., Lovett, G., and Findlay, S., Eds., *Comparative Analysis of Ecosystems. Patterns, Mechanisms and Theories*, Springer-Verlag, New York, 1991.

DeHays, D. H., Winter injury and development of cold tolerance of red spruce, in *Ecology and Decline of Red Spruce in the Eastern United States*, Eagar, C. and Adams, M. B., Eds. Springer-Verlag, New York, 1992, pp. 295-337.

Dixon, R. K., Meldahl, R. S., Ruark, G. A., and Warren, W. G., Eds., *Process Modeling of Forest Growth Responses to Environmental Stress*, Timber Press, Inc., Portland, 1990.

Eagar, C. and Adams, M. B., Eds., *Ecology and Decline of Red Spruce in the Eastern United States*, Ecological Studies No. 96, Springer-Verlag, New York, 1992.

Ennis, C. A., Lazrus, A. L., Kok, G. L., and Zimmerman, P. R., A branch chamber system and techniques for simultaneous pollutant exposure experiments and gaseous flux determinations, *Tellus*, 42B, 170, 1990.

Farman, J. C., Gardiner, B. G., and Shanklin, J. D., Large losses of total ozone in Antarctica reveal seasonal ClO_x/NO_x interaction, *Nature*, 315, 207, 1985.

Fenn M. E. and Bytnerowicz, A., Dry deposition of nitrogen and sulfur to ponderosa and jeffrey pine in the San Bernardino National Forest in Southern California, *Environ. Poll.*, 81, 277, 1993.

Fitzgerald, W. F. and Clarkson, T. W., Mercury and monomethylmercury: present and future concerns, *Environ. Health Persp.*, 96, 159, 1991.

Fowler, D., Cape, J. N., Deans, J. D., Leith, I. D., Murray, M. B., Smith, R. I., Sheppard, L. J., and Unsworth, M. H., Effects of acid mist on the frost hardiness of red spruce seedlings, *New Phytologist*, 113, 321, 1989.

Fox, D. G., Bartuska, A. M., Byrne, J. G., Cowling, E., Fisher, R., Likens, G. E., Lindberg, S. E., Linthurst, R. A., Messer, J., and Nichols, D. S., A screening procedure to evaluate air pollution effects on Class 1 wilderness areas, General Technical Report RM-168, U.S. Department of Agriculture, Forest Service, Rocky Mountain Forest and Range Experiment Station, Fort Collins, CO, 1989.

Friend, A. L., Tomlinson, P. T., Dickson, R. E., O'Neill, E. G., Edwards, N. T., and Taylor, G. E., Jr., Biochemical composition of loblolly pine reflects pollutant exposure, *Tree Physiol.*, 11, 35, 1992.

Gordon, A. G. and Gorham, E., Ecological aspects of air pollution from an iron-sintering plant at Wawa, Ontario, *Can. J. Bot.*, 41, 1063, 1963.

Hallgren, J. E., Linder, S., Richter, A., Troeng, E., and Granat, L., Uptake of sulfur dioxide in shoots of Scots pine: field measurements of net flux of sulphur in relation to stomatal conductance, *Plant Cell Environ.*, 5, 75, 1982.

Hanson, P. J. and Garten, C. T., Deposition of $H^{15}NO_3$ vapor to white oak, red maple and loblolly pine foliage: experimental observations and a generalized model, *New Phytologist*, 122, 329, 1992.

Heagle, A. S., Philbeck, R. B., and Heck, W. W., An open-top chamber to assess the impact of air pollution on plants, *J. Environ. Qual.*, 2, 365, 1973.

Heck, W. W., Adams, R. M., Cure, W. W., Heagle, A. S., Heggestad, H. E., Kohut, R. J., Rawlings, J. O., and Taylor, O. C., Assessing impacts of ozone on agricultural crops. I. Overview, *Environ. Sci. Tech.*, 17, 572A, 1983.

Heck, W. W., Taylor, O. C., and Tingey, D. T., Eds., *Assessment of Crop Loss from Air Pollutants*, Elsevier Applied Science, New York, 1988.

Hendrey, G. R., Lewin, K. F., Kolber, Z., and Evans, L. S., Controlled enrichment system for experimental fumigation of plants in the field with sulfur dioxide, *J. Air Waste Man. Assoc.*, 42, 1324, 1992.

Hogsett, W. E., Olszyk, D., Ormrod, D. P., Taylor, G. E., Jr., and Tingey, D. T., Air pollution exposure systems and experimental protocols, Volume 1: A review and evaluation of performance, EPA/600/3-87/037a, September 1987, Washington, D.C., 1987a.

Hogsett, W. E., Olszyk, D., Ormrod, D. P., Taylor, G. E., Jr., and Tingey, D. T., Air pollution exposure systems and experimental protocols, Volume 2: Description of facilities, EPA/600/3-87/037b, September 1987, Washington, D.C., 1987b.

Hogsett, W. E., Tingey, D. T., and Lee, E. H., Ozone exposure indices: concepts for development and evaluation of their use, in *Assessment of Crop Loss from Air Pollutants*, Heck, W. W., Taylor, O. C., and Tingey, D. T., Eds., Elsevier Applied Science, New York, 1988, pp. 107-138.

Hosker, R. P. and Lindberg, S. E., Review article: atmospheric deposition and plant assimilation of airborne gases and particles, *Atmospheric Environment*, 16, 889, 1982.

Johnson, D. W. and Taylor, G. E., Jr., Role of air pollution in forest decline in eastern North America, *Water Air Soil Pollution*, 48, 21, 1989.

Johnson, D. W. and Lindberg, S. E., Eds., *Atmospheric Deposition and Forest Nutrient Cycling: A Synthesis of the Integrated Forest Study*, Ecological Series 91, Springer-Verlag, New York, 1992.

Johnston, J. W., Shriner, D. S., and Abner, C. H., Design and performance of the exposure system for measuring the response of crops to acid rain and gaseous pollutants in the field, *J. Air Poll. Control Assoc.*, 36, 894, 1986.

Jones, C. G. and Coleman, J. S., Biochemical indicators of air pollution effects in trees: Unambiguous signals based on secondary metabolites and nitrogen in fast growing species, in *Biologic Markers of Air Pollution Stress and Damage in Forests*, National Academy Press, Washington, D.C., 1989, pp. 261-274.

Kauppi, P. E., Mielikäinen, K., and Kuusela, K., Biomass and carbon budget of European forests, 1971 to 1990, *Science*, 256, 70, 1992.

Keeling, C. D., Bacastow, R. B., Carter, A. F., Piper, S. C., and Whorf, T. P., A three-dimensional model of atmospheric CO_2 transport based on observed winds: 1. Analysis of observational data, *Geophys. Monogr.*, 55, 165, 1989.

Kiester, R., Development and use of tree and forest response models, NAPAP Report 17, U.S. National Acidic Precipitation Assessment Program, Washington, D.C., 1990.

Last, F. T. and Watling, R., Eds., *Acidic Deposition: Its Nature and Impacts*, The Royal Society of Edinburgh, Edinburgh, U. K., 1991.

Laurence, J. A., Amundson, R. G., Friend, A. L., Pell, E. J., and Temple, P. J., Allocation of carbon under stress: an analysis of the ROPIS experiments, *J. Environ. Qual.*, 24, 412, 1994.

Lindberg, S. E., Lovett, G. M., Richter, D. R., and Johnson, D., Atmospheric deposition and canopy interaction of major ions in forest, *Science*, 231, 141, 1986.

Lindberg, S. E., Meyers, T. P., Taylor, G. E., Jr., Turner, R. R., and Schroeder, W. H., Atmosphere/surface exchange of mercury in a forest: results of modeling and gradient approaches, *J. Geophys. Res.*, 97, 2519, 1992.

Liu, S., Munson, R., Johnson, D., Gherini, S., Summers, K., Hudson, R., Wilkerson, K., and Pitelka, L., Application of a nutrient cycling model (NuCM) to a northern mixed hardwood and a southern coniferous forest, *Tree Physiol.*, 9, 1783, 1991.

Lovett, G. M., Atmospheric deposition of nutrients and pollutants in North America: an ecological perspective, *Ecol. Appl.*, 4, 629, 1994.

McLaughlin, S. B., Effects of air pollution on forests: a critical review, *J. Air Poll. Control Assoc.*, 35, 512, 1985.

McLaughlin, S. B., and Norby, R. J., Atmospheric pollution and terrestrial vegetation: evidence of changes, linkages, and significance to selection processes, in *Ecological Genetics and Air Pollution*, Taylor, G. E., Jr., Pitelka, L. F., and Clegg, M. T., Eds., Springer-Verlag, New York, 1991, pp. 61-102.

McLaughlin, S. B., Downing, D. J., Blasing, T. J., Cook, E. R., and Adams, H. S., An analysis of climate and competition as contributors to decline of red spruce in high elevation Appalachian forest of the Eastern United States, *Oecologia*, 72, 487, 1987.

McLeod, A. R., Fackrell, J. E., and Alexander, K., Open-air fumigation of field crops: criteria and design for a new experimental system, *Atmos. Environ.*, 19, 1639, 1985.

Miller, P. R., Oxidant-induced community change in a mixed conifer forest, *Adv. Chem.*, 122, 101, 1973.

Mooney, H. A., Winner, W. E., and Pell, E. J., Eds., *Response of Plants to Multiple Stresses*, Academic Press, New York, 1991.

National Academy of Sciences, *Biologic Markers of Air-Pollution Stress and Damage in Forests*, National Academy Press, Washington, D.C., 1989.

Norman, J. M. and Welles, J. M., Radiative transfer in an array of canopies, *Agron. J.*, 75, 481, 1983.

Olson, R. K., Binkley, D., and Böhm, M., Eds., *The Response of Western Forests to Air Pollution*, Ecological Studies No. 97, Springer-Verlag, New York, 1992.

Olszyk, D. M., Takemoto, B. K., Kats, G., Dawson, P. J., Morrison, C. L., Preston, J. W., and Thompson, C. R., Effects of open-top chambers on 'Valencia' orange trees, *J. Environ. Qual.*, 21, 128, 1992.

Pell, E. J., Temple, P. J., Friend, A. L., Mooney, H. A., and Winner, W. E., Compensation as a plant response to ozone and associated stresses: an analysis of ROPIS experiments, *J. Environ. Qual.*, 23, 429, 1994.

Peterson, D. L., Arbaugh, M. J., and Robinson, L. J., Effects of ozone and climate on ponderosa pine (*Pinus Ponderosa*) growth in the Colorado Rocky Mountains, *Can. J. For. Res.*, 23, 1750, 1993.

Reich, P. B. and Amundson, R. G., Ambient levels of ozone reduce net photosynthesis in tree and crop species, *Science*, 230, 566, 1985.

Rogers, H. H., Jeffries, H. E., Stahel, E. P., Heck, W. W., Ripperton, L. A., and Witherspoon, A. M., Measuring air pollutant uptake by plants: a direct kinetic technique, *J. Air Poll. Control Assoc.*, 27, 1192, 1977.

Schroeder, W. H. and Lane, D. A., The fate of toxic airborne pollutants, *Environ. Sci. Tech.*, 22, 240, 1988.

Schulze, E. D., Air pollution and forest decline in Spruce (*Picea abies*) forest, *Science*, 244, 776, 1989.

Schulze, E. D., Lange, O. L., and Oren, R., Eds., *Forest Decline and Air Pollution*, Ecological Studies 77, Springer-Verlag, New York, 1989.

Shaver, C. L., Tonnessen, K., and Maniero, T., Clearing the air at the Great Smoky Mountain National Park, *Ecol. Appl.*, 4, 690, 1994.

Shugart, H. H., Smith, T. M., and Post, W. M., The potential for application of individual-based simulation models for assessing the effects of global change, *Ann. Rev. Ecol. Sys.*, 23, 15, 1992.

Skeffington, R. A. and Roberts, T. M., The effects of ozone and acid mist on Scots pine seedlings, *Oecologia*, 65, 201, 1985.

Smith, W. H., *Air Pollution and Forests*, Springer-Verlag, New York, 1990.

Swain, E. B., Engstrom, D. R., Brigham, M. E., Henning, T. A., and Brezonik, P. L., Increasing rates of atmospheric mercury deposition in midcontinental North America, *Science*, 257, 784, 1992.

Tandy, N. E., Di Giulio, R. T., and Richardson, C. J., Assay and electrophoresis of superoxide dismutase from red spruce (*Picea rubens* Sarg.), loblolly pine (*Pinus taeda* L.) and Scotch pine (*Pinus sylvestris* L.), *Plant Physiol.*, 90, 742, 1989.

Taylor, G. E., Jr., Johnson, D. J., and Andersen, C. P., Air pollution and forest ecosystems: a regional to global perspective, *Ecol. Appl.*, 4, 662, 1994.

Taylor, G. E., Jr., and Hanson, P. J., Forest trees and tropospheric ozone: role of canopy deposition and leaf uptake in developing exposure-response relationships, *Agr. Ecosys. Environ.*, 42, 255, 1992.

Temple, P. J., Reichers, G. H., Miller, P. R., and Lennox, R. W., Growth responses of ponderosa pine to long-term exposure to ozone, wet and dry acidic deposition, and drought, *Can. J. For. Res.*, 23, 59, 1993.

Teskey, R. O., Dougherty, P. M., and Wiselogel, A. E., Design and performance of branch chambers suitable for long-term ozone fumigation of foliage in large trees, *J. Environ. Qual.*, 20, 591, 1991.

Thompson, A. M., The oxidizing capacity of the earth's atmosphere: probable past and future changes, *Science*, 256, 1157, 1992.

Tingey, D. T., Hogsett, W. E., and Henderson, S., Definition of adverse effects for the purpose of establishing secondary national ambient air quality standards, *J. Environ. Qual.*, 19, 635, 1990.

Tjoelker, M. G., and Luxmoore, R. J., Soil nitrogen and chronic ozone stress influence physiology, growth and nutrient status of *Pinus taeda* L. and *Liriodendron tulipifera* L. seedlings, *New Phytologist*, 119, 69, 1991.

Tjoelker, M. G., Volin, J. C., Oleksyn, J., and Reich, P. B., An open-air system for exposing forest canopy branches to ozone pollution, *Plant Cell Environ.*, 17, 211, 1994.

Travis, C. C., and Hester, S. T., Global chemical pollution, *Environ. Sci. Tech.*, 25, 814, 1991.

Ulrich, B., Production and consumption of hydrogen ions in the ecosphere, in *Effects of Acid Precipitation on Terrestrial Ecosystems*, Hutchinson, T. C. and Havas, M., Eds., Plenum Press, New York, 1980, pp. 255-282.

Urban, D. L., Bonan, G. B., Smith, T. M., and Shugart, H. H., Spatial applications of gap models, *Forest Ecol. Man.*, 42, 95, 1991.

Voglemann, J. E., Detection of forest change in the Green Mountains of Vermont using multispectral scanner data, *Int. J. Rem. Sens.*, 9, 1187, 1988.

Volz, A. and Kley, D., Evaluation of the Montsouris series of ozone measurements made in the nineteenth century, *Nature*, 332, 240, 1988.

Weinstein, D. E. and Yanai, R. D., Integrating the effects of simultaneous multiple stresses on plants using the simulation model TREGRO, *J. Environ. Qual.*, 23, 418, 1994.

Weinstein, D. A., Beloin, R. M., and Yanai, R. D., Modeling changes in red spruce carbon balance and allocation in response to interacting ozone and nutrient stresses, *Tree Physiol.*, 9, 127, 1991.

West, D. C., McLaughlin, S. B., and Shugart, H. H., Simulated forest response to chronic air pollution stress, *J. Environ. Qual.*, 9, 43, 1980.

Westman, W. E., Oxidant effects on California coastal sage scrub, *Science*, 205, 1001, 1979.

Zillioux, E. J., Porcella, D. B., and Benoit, J. M., Mercury cycling and effects in freshwater wetland ecosystems, *Environ. Toxicol. Chem.*, 12, 2245, 1993.

5 Multiple Environmental Stresses on the Fragile Lake Tahoe Ecosystem

Charles R. Goldman

Lake Tahoe, revered for its cobalt-blue color and remarkable transparency, is a system undergoing multiple environmental stresses. An aerial photograph (Figure 5.1) shows Tahoe from an altitude of 19,000 feet and its watershed, which is very small (800 km^2) with respect to its surface area (500 km^2) and volume (156 km^3). It owes its low fertility to this small watershed and the relatively infertile soils of the basin. During the past century, populations have expanded dramatically along the shores of most of the world's lakes, whose watersheds have often been drastically changed from virgin forests to agricultural fields, urban areas, or industrial complexes. Tahoe is no exception. Since its discovery by John Fremont in 1844, the basin has been lumbered for mining timber and undergone extensive development as the recreational values of the Tahoe basin have become better known. The road network has expanded, wetlands have been destroyed, and nutrient loading from the streams and atmosphere has increased. As a result, there has been a steady, well-documented increase in the fertility of Tahoe (Figure 5.2) which continues to reduce the remarkable transparency of the lake (Figure 5.3). In comparison to most of the world's lakes, Tahoe is still in very good condition. Lake Baikal in Russian Siberia, which is often compared with Tahoe, is far less transparent throughout most of the year. During the month of August 1993, for example, we never saw a transparency greater than eight meters and generally it was less than four meters. Baikal, during its ice-free period, is a much more productive system than Tahoe and has some areas which are extremely productive.

While discussing environmental quality, we must keep in mind that people vary in their perception of the aesthetic qualities that make any given environmental situation more desirable than another. Nowhere do we have any better evidence of this than the contrast between ultraoligotrophic Lake Tahoe and hypereutrophic Clear Lake, in California. The public would very much like Clear Lake to look more like its name or, at the very least, be free of the serious blue-green algal blooms that give it the appearance of pea soup. In the meantime, Clear Lake continues to produce a lot of fish. Unfortunately, those high in the food chain have a high mercury content. In China, lakes like Clear Lake are shallow, highly productive, and extraordinarily important for the fish protein they produce. The last thing the Chinese would want

1-56670-309-3/98/$0.00+$.50

FIGURE 5.1 Aerial photograph of Lake Tahoe looking toward the southeast. The 500-square-km lake is situated in a graben fault at the crest of the Sierra Nevada between Nevada and California.

to do is reduce the lakes' high trophic state causing a corresponding decline in fertility. The large (2425 km^2) Taihu Lake in China near Shanghai averages less than two meters in depth, is extremely eutrophic, and produces enormous quantities of fish and shellfish. In 1984 an intensively managed 25 ha area reported 8250 kg/ha fish production, which in 1985 increased to 15,750 kg/ha (Chang 1994). The water, however, is also used for domestic purposes and large beds of aquatic plants are employed as biological pre-filters to reduce the algal concentrations that must eventually be removed in their treatment plants.

Unlike ecotoxicologists, limnologist-ecologists look first at the effects of environmental stresses and then try to work backward through the system to tease out cause and effect. In the comments that follow, I would like to point out the importance and essential nature of long-term data when attempting to evaluate change in ecological systems as well as the dominance of some climatic and physical processes such as wind, precipitation, and forest or brush fires in environmental analysis. I will underscore the very heavy nitrogen input from the atmosphere at Tahoe. There has been, largely because of this source, a shift from what was a classical nitrogen limitation in the sense of Liebig's "law of the minimum," to a system that is increasingly phosphorus-sensitive. The atmospheric loading of nitrogen to the Tahoe basin has turned out to be much greater than we had imagined. Last, but not least, we will consider lakes as reservoirs of history in the sense that they accumulate in their sediments whatever has occurred on their watershed, and detail in their record that events like drought have occurred repeatedly in the past and have a tremendous impact on lakes.

Without the recent, human-induced impacts on the lake and particularly its highly erodible watershed, the nutrient loading would be very low. The population has grown rapidly in the basin (see Figure 5.2 inset), and with it the problem of sewage treatment and disposal. In the early 1960s and 1970s, the engineering "technological fix" was the popular idea of the day. Tertiary treatment of the sewage within the

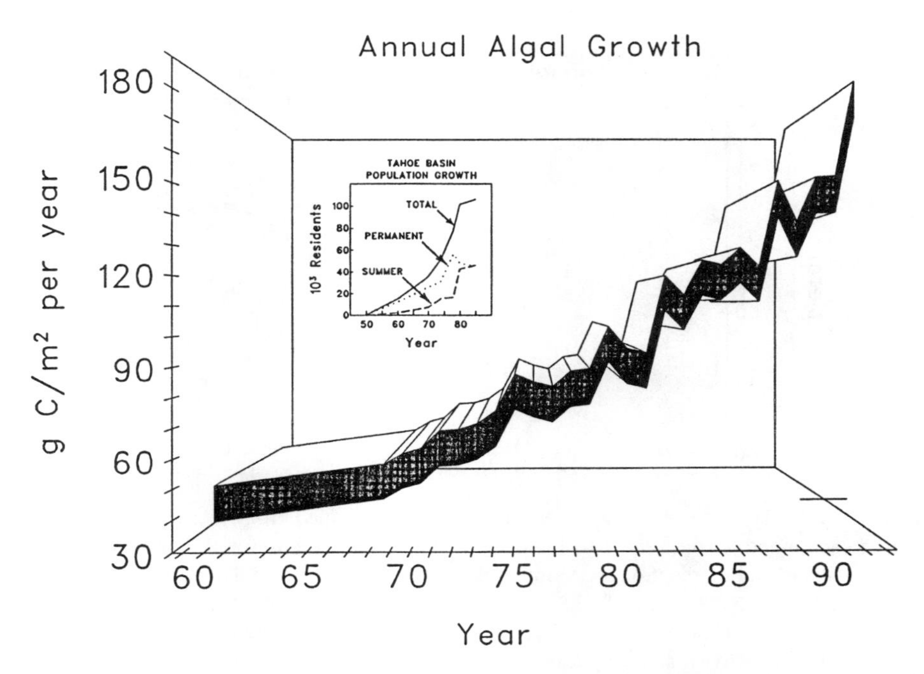

FIGURE 5.2 Annual algal growth rate in grams of carbon per square meter per year based on the integration of approximately 35 measurements taken *in situ* with ^{14}C during each year. Only spotty data are available during the period from 1960 through 1966. Intensive measurements began in 1967 at 13 depths on a weekly or biweekly schedule. *In situ* ^{14}C measurements were first integrated by depth, then the monthly averages integrated to give the annual carbon fixation rate beneath an average square meter of surface area. Algal productivity increased on average 5.5% per year, with most of the interannual variability resulting from the depth of winter mixing. The insert shows the population increase from 1950 to the 1980s.

basin was a favored approach since it was believed by some that the effluent could be returned to the lake. By performing bioassay experiments on the effluent from the South Tahoe Public Utility District (STPUD) plant, I was able to demonstrate how extremely stimulating to algal growth tertiary effluent would be. Although the treatment system was very good at removing phosphorus, this element was in those days less important to remove than nitrogen because the lake was essentially nitrogen-limited. In other words, if you added phosphorus to cultures of lake water during the 1960s, you would actually inhibit algal photosynthesis. If you added nitrogen or, for that matter, tertiary effluent (which was high in ammonia-nitrogen), algal growth was greatly increased. Using the results of these bioassay experiments, I was able to convince a distinguished group of engineers, including Rohlic from Wisconsin, Kauffman from Stanford, and McGauhey and Pearson from Berkeley, that tertiary treatment was ineffective for controlling the eutrophication of Tahoe since the high ammonia concentrations were not adequately reduced by the treatment process. Further proof of this was provided when, after some struggle, they finally agreed to export the sewage out of the basin to Indian Creek Reservoir. As predicted,

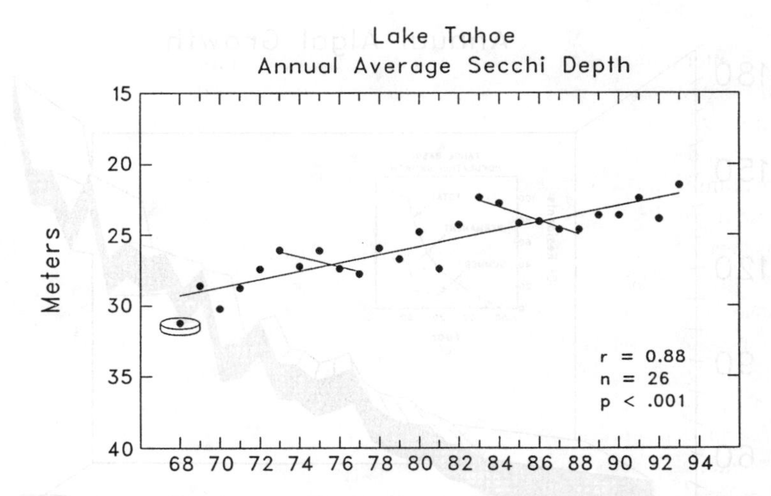

FIGURE 5.3 The annual average Secchi depth is determined from measurements taken during each month of the year. Two lines intersecting the long-term trend show the increase in transparency during the drought period of the 1970s and during the five years immediately following the El Niño event of 1983, when deep mixing did not occur. Transparency is decreasing at approximately a half-meter per year with considerable interannual variability which, like primary productivity, is largely explained by the variability in depth of winter mixing.

the photosynthetic levels of the algae were greatly elevated by the ammonia present in the effluent, turning the reservoir bright green every spring. Extremely toxic ammonium hydroxide was formed as the pH of the system rose toward 10. This periodically killed the trout in the reservoir. Because tertiary treatment is expensive in terms of both energy and chemicals, the STPUD plant has returned to secondary treatment.

It is important to keep in mind that, even in the crudest sense of the word, Tahoe is occasionally polluted by sewage. I was able to follow, during the spring snowmelt period, a trail of toilet paper and fecal material during an overflow or backup from a sewer right down and into the lake. We photographed the event right behind a shopping complex in Tahoe City. At other times, infiltration into jointed sewer lines overwhelms the capacity of the treatment plant, and over-flows of untreated effluent flow downhill to the lake.

Another major perturbation of the Tahoe ecosystem has been the loss of valuable wetlands which, in addition to providing wildlife habitat, serve as sediment settling areas and nutrient sponges for the watershed. The creation of the Tahoe Keys development destroyed a major portion of this wetland filtration capability of the mouth of the Upper Truckee River. This river supplies about 25% of the total stream inflow to Tahoe. This loss of the Pope Marsh was serious, and one can observe in spring the sediment plume being carried into the lake and directed by effects of wind and perhaps Coriolis force east along the less-developed eastern side of the lake.

When I began my studies in 1959 at Tahoe, nobody wanted to believe that Tahoe was actually changing for the worse. Through the late 1960s we were already recording an annual increase in primary productivity (algal growth rate), as measured with the sensitive carbon-14 method, of about 5.5% per year (Figure 5.2). Notice the large interannual fluctuation in primary productivity resulting from variability in such weather effects as temperature and storms. In studies of this kind it is very important, but often difficult or impossible, to have another lake serve as a "control." Since development was proceeding at a high rate in the Tahoe basin, the developers were becoming more and more sophisticated as we entered the 1970s and 1980s, hiring better and more-experienced consultants, and forming their own pro-development group named the Tahoe Sierra Preservation Council. Recalling the extensive timber removal from the basin during the 1860s to shore up the mines of the Comstock Lode in Nevada and also noting the eutrophication of many eastern lakes, they asked, "Why worry about Tahoe's increase in fertility? After all, it is a general phenomenon of lakes worldwide, and therefore should not be attributed to development of the basin." Fortunately, we had simultaneously begun conducting exactly the same primary productivity, transparency, and nutrient studies at undeveloped Castle Lake, which is located in the Klamath Mountains just to the west of Mt. Shasta at approximately the same elevation as Tahoe. By comparing the primary production of the two lakes, we could demonstrate that at Lake Tahoe there has been a steady and significant increase since 1959, while at Castle Lake, which like Tahoe shows tremendous interannual variability, there had been no consistent upward trend.

I was, of course, accused of being an extreme environmentalist by the vocal opposition to regulating growth in the basin headed by the development forces and some of the real estate concerns. People began to believe me only after Tahoe began to develop a green margin from increasing growth of periphyton (attached algae), which appeared in prodigious quantities around the perimeter of the lake. The largely shore-bound public could actually see this green rim develop each spring and slowly began to understand that there was something to be very concerned about. In those days, the growing membership and campaign of the League to Save Lake Tahoe were largely responsible for educating the public. The Tahoe Research Group produced and distributed three documentary films to aid in this effort.

Despite what has been reported above, it is important to keep in mind that Tahoe is still remarkably transparent among the world's large lakes and has only just entered the earliest stages of eutrophication. This carefully documented increase in fertility and loss of transparency is to a large extent due to atmospheric inputs of nitrogen.

The Secchi disk, with no moving parts, is by far the cheapest limnological instrument in existence. It consists of a 20-cm flat white disk which is lowered into the water until it disappears, then pulled back up again until it reappears. The average of the two measurements is termed the Secchi depth. The observer is viewing a light path in Tahoe down to the disk and back again which can be as great as 50 to 80 meters. Since a typical photometer uses only a one-to-ten centimeter path length, one can appreciate that the observer actually gets a very accurate measure of transparency with the Secchi disk. It was developed by the Italian astronomer and Jesuit priest Pietro Angelo Secchi in 1865 for the papal ship *S.S. l'Immacolata Concezione* for coastal Mediterranean waters. Since then it has had wide application in both limnology and oceanography.

Examination of the Secchi depth data shows that we are losing approximately half a meter of transparency per year; this regression, based on about 35 measurements per year for more than 27 years, is significant at the 0.001 level (Figure 5.3). Note, however, that there are some extreme reversals in this long-term trend, which are another reminder of the great importance of long-term data in environmental research. If, for instance, we had evaluated the transparency of Tahoe based only upon the 1973 to 1977 or 1983 to 1988 data, which might have been collected during a typical three, four, or possibly even six-year research grant, we would have erroneously concluded that Tahoe was actually becoming more transparent and oligotrophic. The short regressions drawn on Figure 5.3 correspond to these two drought periods. This illustrates the inherent danger of using short-term data sets to assess environmental conditions since, during both of the recent drought periods in California, Tahoe actually regained some lost transparency. This improved transparency during the drought years resulted from a lack of late-winter storms capable of mixing the lake to depths greater than about 250 m where nitrogen concentrations are higher. In the early days of our studies, we focused very heavily on stream nutrient discharge from the watershed. When the total stream discharge of nitrate was plotted against algal growth rate, we found that, if we eliminated the heavy runoff year of 1982 which was a real outlier in the data set, we obtained a good correlation between the two. In fact, by omitting that one data point, the level of significance improved from 0.1 to 0.01.

The investigator should carefully consider cause and effect when an outlier of this sort appears among an otherwise good fit for a regression line. By comparing the outlier data of 1982 with 1983, it was obvious that very similar stream discharges had produced very different annual productivity. After close examination of the data, I discovered that in 1983 the lake mixed all the way to the bottom, but in 1982, despite heavy precipitation, there were no winter winds sufficiently strong to mix the lake to the bottom. We can actually determine the depth of mixing by analyzing the nitricline formed by removal of nitrogen by phytoplankton near the surface. During mixing, this feature becomes less prominent and sometimes disappears down to the depth of the mix (Figure 5.4). Phosphorus remains rather uniform, except where it increases slightly just above the bottom. Nitrate, however, accumulates in the deep waters for several years if the lake does not undergo a complete winter mix. In fact, Tahoe tends to mix completely only every 3 to 4 years. Going into February 1983, deep-water nitrate levels were much higher than those nearer the surface. But by 31 March, the lake had mixed to the bottom, resulting in a uniform distribution of nitrate throughout the water column. Deep mixing following one or more years of partial mixing stimulates algal growth by bringing a large influx of nitrate into the lighted portion of the water column, called the euphotic zone. In 1983, when the lake mixed to 450 meters, the annual algal growth rate increased 18% over 1982. Typical drought-type conditions then developed with no late winter storm in 1984 and a spring mixing depth of only 150 meters.

A modeler's dream, the sufficiently long-term Tahoe data set is ideal to work with. If a plot (Figure 5.5) is made of primary production vs. the depth of mixing, more than 64% of the variance is accounted for (Goldman and Jassby 1990). In fact, one scarcely needs to model data that provides such clear, irrefutable evidence. Since

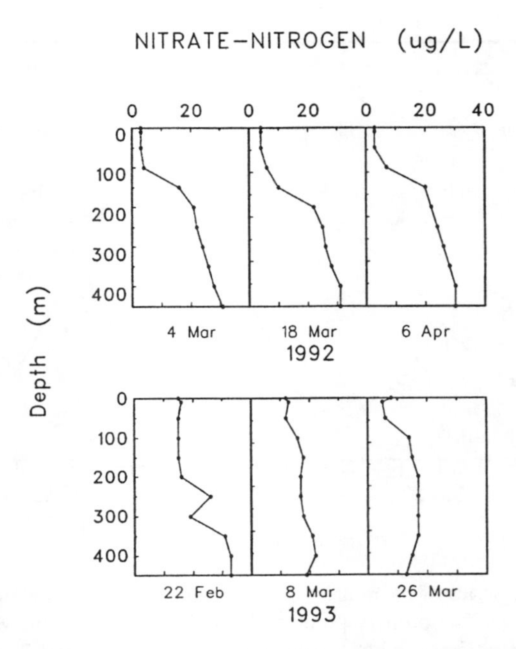

FIGURE 5.4 Nitrogen distribution in the Lake Tahoe water column taken at the mid-lake station in March and April of 1992 and in February and March of 1993. Note that mixing during 1992 extended only to approximately 200 meters, whereas in 1993 there was a complete mix redistributing the nitrogen and other algal nutrients throughout the water column. The nitricline provided the strongest evidence of the depth of mixing and was dependent upon the evolution of sensitive nitrate–nitrogen analytical methodology. Highest rates of productivity occur during years like 1983 when nutrients trapped in the deep water of the lake are brought to the surface through complete mixing.

I began these studies at Tahoe in 1959 there has been a great deal of atmospheric loading of nitrogen, resulting in a distinct shift from nitrogen limitation in the lake to increasing phosphorus sensitivity. This steady nitrogen loading from the atmosphere has exceeded the annual stream nitrogen inputs. At Tahoe we have witnessed a shift from a typically nitrogen-limited system during the 1960s through the early 1980s, to a lake where nitrogen is now considerably less stimulating than phosphorus. Most lakes are more likely to be co-limited by both nitrogen and phosphorus (Elser et al. 1990). For example, if more phosphorus is added to Tahoe, one can get greater stimulation by adding nitrogen as well.

If we look at the long-term nitrogen data, we find that there has been a steady accumulation of nitrogen in the system as evidenced by an increase in the nitrogen-to-phosphorus ratio. To evaluate better atmospheric loading, it was necessary to set up a series of stations both on the lake and near the shore. Examination of wet/dry precipitation collectors to assess the relative inputs of nutrients found in dust, snow, and rain revealed the heaviest atmospheric deposition at an upper elevation site in Ward Valley on the west side of the lake, an appreciably reduced input at lake level, and then considerable variation eastward onto the lake (Figure 5.6; Jassby et al.

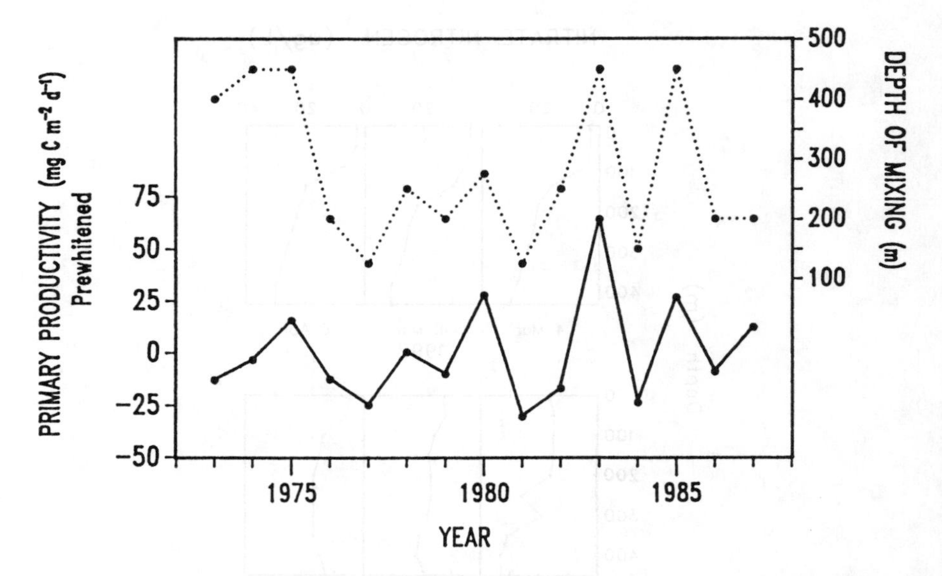

FIGURE 5.5 Time series of mean daily primary productivity and maximal depth of spring mixing at Lake Tahoe. The primary productivity values have been prewhitened by filtering with a Markov process (From Goldman, C. R., Jassby, A., and Powell, T., *Limnol. Oceanogr.*, 34(2), 310, 1989. With permission.)

1994). The midlake station, located on a spar buoy, tended to have the lowest level of atmospheric deposition.

As noted earlier, drought is an important feature of our west coast. Surveys of the bottom of the lake with our remote-operated underwater vehicle (ROV) have revealed tree trunks from a forest that existed some 5000 years ago, indicating that lake levels were then several meters lower than during the current drought. Since the trees were over 100 years old when they died, this was a long and much more serious drought than we have encountered during modern history. With the ROV operated from aboard our research vessel, we were able to locate and explore the steamship *Tahoe,* which was scuttled off Deadman's Point in 1942. In the washroom's wash basin, we can clearly see the accumulation of sediments which add about 2 mm per year to the bottom of Tahoe. We are currently reconstructing the last few centuries by examining sediment cores taken from various locations in the lake. We have already observed a drastic change in the diatom species assemblage since intensive development of the basin began in the late 1950s (Byron and Eloranta 1984).

To conclude, the current research projects at Tahoe involve continued attention to the multiple stresses on Lake Tahoe and its watershed. New analysis of the deep sediments to assess the historical sedimentation rate and sediment contents is under way. Our analysis of sediment and nutrient transport to the lake will be related to urban, geologic, and physiographic characteristics of the individual watersheds, particularly in relation to possible climate change. Special attention will be given to the relative importance of watershed vs. atmospheric sources of nutrients. Now

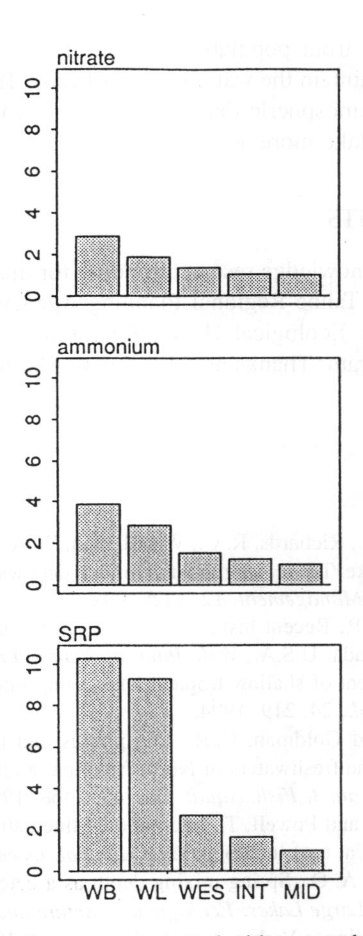

FIGURE 5.6 Relative deposition rates measured with snow tubes for sampling on land and on spar buoys on the lake covering the period November 17, 1987, to February 7, 1992. WB = Ward Valley Bench; WL = Ward Lake Level; WES = West Lake; INT = Intermediate Lake; MID = Mid Lake (From Jassby, A. D., Reuter, J. E., Axler, R. P., Goldman, C. R., and Hackley, S. H., *Water Resources Research,* 30, 2207, 1994. With permission.)

that phosphorus has become such an important limiting factor for algal growth in the lake, watersheds, which remain the main source of phosphorus input, must be managed more carefully than they were in the 1960s when nitrogen was the nutrient most responsible for eutrophication of the lake. Quantification of biologically available phosphorus (BAP) from tributary waters and evaluation of erosion control effectiveness are both of particular importance. Although millions of dollars are spent on building and maintaining various erosion control structures and settling basins, there has to date been very little effort to evaluate the relative effectiveness of the different management practices. The feasibility of using water runoff detention basins, artificial wetlands (Reuter et al. 1992), and other methods to ameliorate the effects of erosion is in dire need of further scientific evaluation. The ecology and reproduction of the lake trout population are also under study (Beauchamp et al. 1992). To protect and maintain the water quality of Lake Tahoe, an empirical model linking watershed and atmospheric deposition processes is needed to evaluate the multiple stresses on the lake more accurately.

ACKNOWLEDGMENTS

The author wishes to acknowledge research funding for this work from the National Science Foundation, the Tahoe Regional Planning Agency, the University of California, Davis, Center for Ecological Health Research, and the Lake Tahoe Interagency Monitoring Program. Thanks are also due to Ms. Arneson, who helped edit the manuscript.

REFERENCES

Beauchamp, A., Allen, B. C., Richards, R. C., Wurtsbaugh, W. A., and Goldman, C. R., Lake trout spawning in Lake Tahoe: egg incubation in deepwater macrophyte beds, *No. American J. Fisheries Management*, 12, 442, 1992.

Byron, E. R. and Eloranta, P., Recent historical changes in the diatom community of Lake Tahoe, California–Nevada, U.S.A., *Verh. Internat. Verein. Limnol.*, 22, 1372, 1984.

Chang, W. Y. B., Management of shallow tropical lakes using integrated lake farming, *Verh. Internat. Verein. Limnol.*, 24, 219, 1994.

Elser, J. J., Marzolf, E., and Goldman, C. R., Phosphorus and nitrogen limitation of phytoplankton growth in the freshwaters of North America: a review and critique of experimental enrichments, *Can. J. Fish. Aquat. Sci.*, 47, 1468, 1990.

Goldman, C. R., Jassby, A., and Powell, T., International fluctuations in primary production: Meteorological forcing at two subalpine lakes, *Limnol. Oceanogr.*, 34(2), 310, 1989.

Goldman, C. R. and Jassby, A. D., Spring mixing depth as a determinant of annual primary production in lakes, in *Large Lakes: Ecological Structure and Function*, Serruya, C. and Tilzer, M. M., Eds., Springer-Verlag, New York, 1990, pp. 125-132.

Jassby, A. D., Reuter, J. E., Axler, R. P., Goldman, C. R., and Hackley, S. H., Atmospheric deposition of nitrogen and phosphorus in the annual nutrient load of Lake Tahoe (California–Nevada), *Water Resources Research*, 30, 2207, 1994.

Reuter, J. E., Djohan, T., and Goldman, C. R., The use of wetlands for nutrient removal from surface runoff in a cold climate region of California — Results from a newly constructed wetland at Lake Tahoe, *J. Environ. Man.*, 36, 35, 1992.

Part Two

Establishing the Health of Ecosystems

6 Ecosystem Health: Some Perspectives

Barry W. Wilson

"Chasing definitions is a harmless vice, like eating peanuts...but...don't overdo it."

Once, in those halcyon days when environmental courses were novelties in a technical curriculum, I accompanied a student to the outskirts of town to help her with a project on burrowing owls. In the hope of finding them in their burrows, I gravely knelt by the side of the road and pressed a stethoscope to the ground. Soon, I noticed cars slowing down to watch us, and I heard a little girl ask, "Mommy, is that man listening to the heartbeat of the Earth?" Would it were that simple to determine the health of the planet. We Earth doctors could give the planet an environmental aspirin and ask it to call us in several millennia if it did not feel better.

The burrowing owls have long since departed. They were displaced by ranch-style homes where there are no longer ranches, but the problem of assessing the health of ecosystems remains. Unless, of course, one takes the position that the concept of "health" is inappropriate above the organism level. One argument is that it smacks of a kind of "pop ecology," applying simple paradigms to complicated situations. Another is that populations that have survived the ministrations of humans are "healthy" by virtue of their very existence. Nevertheless, biologists from many disciplines are both concerned about and intrigued by the idea that the health of environments can be both defined and assessed.

One must admit that at this stage of our knowledge of the interrelationships and rate-limiting processes in plant and animal communities, raising the subject of health as an ecological paradigm is likely to lead to more questions than answers (Table 6.1). One basic question is whether ecosystem health is a subjective concept that, like beauty, is in the eye and mind of the beholder, a metaphor for all the inquirer holds near and dear, or whether it possesses objective, determinable properties. Is it a matter of the stability of species over time in an ecosystem, or of their normal progression in the face of orderly changes in habitats? Is it a question of establishing the sustained productivity of a habitat, of inquiring whether the biomass is remaining within the carrying capacity of the land? Has the notion of ecosystem health got something to do with numerical indices of reproductive fitness? Politically, how is it related to socioeconomic values? Is an ecosystem useful to humans "healthy" even though it may be sustained by management methods that are energetically inefficient and biologically restrictive? In a world more and more driven by environmental impact assessments, can "health" be used as an indicator, indeed,

1-56670-309-3/98/$0.00+$.50

TABLE 6.1
Ecosystem Health: Paradigm or Metaphor?

In the eye and mind of the beholder?
Species stability, normal progression?
Remaining within carrying capacity? Sustained productivity?
Optimal reproductive fitness?
Attaining socioeconomic values?
Quantitative risk assessment indicator?
Definable on ecological, physiological thermodynamic principles?
Ill-definable as human health?

TABLE 6.2
Ecosystem Health Paradigms and the Hierarchy of Life

Ecosystem/Community: Optimal biodiversity of native species, trophic levels
Population: Sustainable or natural trajectory, genetic diversity
Individual: Optimal reproduction and other physiological states
Cell/Organ: Absence of overt pathology
Thermodynamic: Relative minimization of entropy production

as a driver for quantitative risk assessment? Biologically, can ecosystem health be defined on ecological, physiological, even on thermodynamic grounds?

What about the concept of human health, given that the idea of ecosystem health is derived from it by analogy. How clearly can one define human health? Is it merely the absence of overt disease? Is it the response of an individual to insult, injury, or stress within what is accepted as normal physiological limits? Indeed, in the rapidly moving world of molecular biology, is it a matter of having the right genes in the right place at the right time?

Perhaps there are different acceptable definitions for health at each level of the hierarchy of life (Table 6.2). Optimal biodiversity and existence of trophic levels could apply to an ecosystem or an individual habitat. Sustainability or progression along a natural trajectory and optimal genetic diversity might apply to populations of a single species. Optimal reproduction and maintenance of a normal physiological state could be applied to the individual animal, plant, or protistan; the absence of overt pathology would be suitable for the organ and cellular levels, and the relative minimization of entropy production would be a characteristic of "health" on the thermodynamic level. A socioeconomic approach might define a healthy system as one that maximizes human benefits without sacrificing any of the above features of the environment and its biota.

How ecosystem health can be studied is, of course, dependent on the brand of health one is touting. But there may be some common features (Table 6.3) including: (1) positive feedback to stresses involving relatively rapid recoveries, (2) maximum

TABLE 6.3
Characteristics of Healthy Systems

Positive feedback to stresses, relatively rapid recoveries
Maximum biodiversity of native species, occupied trophic levels
Sustainable reproduction rates
Minimal pathology
Genetic diversity
Time frames: minutes to millennia

bio- and genetic diversity on an ecosystem level, (3) sustainable reproduction rates, and (4) minimum pathogenic impacts on population and individual levels within time frames of minutes to millennia depending on the systems concerned.

The authors of this section on ecosystem health do not present a common position on establishing the wellness or study of the specific components of ecosystems. Issues discussed in some detail include: whether the health of an ecosystem can be strictly defined at all or whether we will have to be satisfied with empirical definitions, describing the success or failure of particular species or the future development of an entire ecosystem along appropriate and productive evolutionary trajectories. The appropriate role of physiological and biochemical biomarkers in the study of multiple stressors and their importance in present and future environmental management decisions is discussed, as is the relationship between establishing the health of an ecosystem and its components and the rapidly emerging area of environmental risk assessment — characterizing and managing risks posed by single and multiple stresses.

Alan Heath opens the section with a discussion of the physiology of fish and their usefulness in studying the health of ecosystems. Pierre Mineau focuses on biochemical and physiological biomarkers; he discusses pesticides and songbirds, using cholinesterases, a favorite class of enzymes of mine, with the actions of fenitrothion as examples. Anne Fairbrother notes the "great allure" held by the health paradigm to environmental scientists. She presents a case for the indefinableness of ecosystem health except as a framework for discussion. Fairbrother recommends it be replaced by risk characterization, establishing the consequences of anthropogenic stressors. Tom McKone presents a useful risk assessment primer — its relationships to environmental health and to risk management. Ed Goldberg examines cause and effect relationships of single and multiple stressors in the marine environment. And finally, Bill Lasley offers a parallel with human reproductive function.

Metaphor or method? The subject of ecosystem health is not laid to rest by the authors in this section, but the issues they raise explore some of the length and breadth of the subject. A pessimist might harken to the warning of Alan Heath, "If you don't know what you are looking for, you are not likely to find it!", but an optimist may take hope from the thought that it is better to do a hundred experiments to seek truth than none and be satisfied with error. Given the complexity of the webs of life within ecosystems, personally, I think it unlikely there will be one parameter agreeable to everyone as a satisfactory indicator of the health, the heartbeat of an ecosystem. One reason is the lack of quantifiable one-to-one relationships between the success of an individual species and the ecosystem in which it lives. The whole is indeed more than the sum of its parts. Regardless, studies on the molecular, cellular, organ, and organism levels can document the adverse effects of single and multiple stressors in the environment, and serve as "early warning signs" of harm being done at some level to the ecosystem itself.

With a tip of the hat to Barry Commoner and Ian Nisbet, I gently remind the reader of several important matters concerning ecosystems, statements that, in their own way, resemble the Laws of Thermodynamics to which we are subject (Table 6.4).

TABLE 6.4
Some Ecological Principles

Everything is connected to everything else
Everything must go someplace
There is no free lunch

A fourth point might be to recognize that, as an experimental paradigm, death is a lousy endpoint! Whatever we espouse as principles, it is important to subject them to the harsh test of field study. Whether environmental or human oriented, in the long run, definitions of health will be as useful as their ability to be critically tested.

Finally, one mark of credibility of a concept is its recognition within a granting agency (e.g., molecular biology in the National Science Foundation) or by a publisher. In this case David Rapport and his colleagues have successfully given birth to a journal of *Ecosystem Health,* and the term is used in the title of other publications. Rapport's editorial in volume 2, # 2 of the journal is entitled "Encouraging Dissent"; he notes that "without vigorous challenges to goals, concepts, and methods, no discipline or transdiscipline can flower." It is safe to say that, given the extent of discourse on the topic, the area of ecosystem health will surely bloom.

The announcement of the conference upon which this volume is based closed with homage to far-seeing ecologist Aldo Leopold, who developed the idea of a "Land Ethic." I would like to believe he would not bridle at including health as one of its components, and that he would be willing to take a holistic approach to understanding the living landscape. In the announcement, I called attention to Leopold's essay on the Gavilan River. In it he likened an ecosystem to a great orchestra with each plant and animal as one of its instruments. Leopold chided scientific specialists (like myself, and maybe you too) for listening only to the instrument they study, ultimately dismembering it to learn about it, and finally destroying the orchestra itself. The chapters that follow affirm the importance of the music inherent in multidisciplinary research and the uniqueness of the knowledge gained through synthesis rather than dissection.

SELECTED REFERENCES

Leopold, A., *A Sand County Almanac*, Oxford University Press, Oxford, U.K., 1949.

Rapport, D. J., Encouraging dissent, *Ecosystem Health*, 2(2), 99, 1996.

Cairns, J., Jr., and Niederlehner, B. R., Ecosystem health as a management tool, *J. Aquatic Ecosystem Health*, 4(2), 91, 1995.

Cairns, J., Jr., and Niederlehner, B. R., Predictive ecotoxicology, in *Handbook of Ecotoxicology*, Hoffman, D. J., Rattner, B. A., Burton, G. A., Jr., and Cairns, J., Jr., Eds., Lewis Publishers, Boca Raton, FL, 1995, 667.

Epstein, P. R., Emerging diseases and ecosystem instability: new threats to public health, *Am. J. of Public Health*, 85(2), 168, 1995.

7 Physiology and Ecological Health

Alan G. Heath

INTRODUCTION

Physiology has had a long tradition of being intimately involved with the assessment of the health of humans. As part of a physical exam the doctor measures blood pressure, serum cholesterol, and hemoglobin concentration to evaluate our health. Our body experiences multiple stresses caused by job pressures, poor diet, disease organisms, and environmental pollutants, all of which affect our physiology. But health of an ecosystem? Some might hope for some physiological measurements one could make of an ecosystem's health, a sort of physical that your friendly family ecological physiologist might perform. I am afraid that is wishful thinking, as attractive as it appears. Indeed, there are some who argue that there is no such thing as ecosystem health. Suter (1993) maintains that the term "health" is a metaphor that is misleading when applied to ecosystems because they are not like an organism with boundaries and homeostatic controls. I do not intend to discuss the pros and cons of that argument. Rather, the theme I wish to develop is that the organisms in an ecosystem do have boundaries and homeostatic controls and these are affected by multiple stresses in the ecosystem.

The physical and chemical conditions (e.g., temperature and dissolved oxygen) of an ecosystem are constantly changing, and contaminant stresses from inputs of pollutants often get superimposed on these natural stressors. By measuring the physiological condition of the animals, one can obtain an indirect measure of the health (or quality) of the ecosystem they are living in. The theory is that the organisms will integrate the impact of these multiple stresses and thus serve as a sort of sensor, a sensor that has the advantage over chemical and physical measurements alone because it tells us whether the environmental stresses are doing any actual harm to the residents. And many of these measurements can tell us if the organisms are in trouble long before the population declines or the ecosystem becomes badly degraded.

My orientation here is aquatic ecosystems, and the organisms I will be emphasizing are fish, although the same principles could also be applied to large invertebrates. Fish make good sentinels in part because they are usually toward or at the top of the food chain and thus, an ecosystem that is maintaining healthy diverse fish populations is probably in good condition (Fausch et al. 1990).

1-56670-309-3/98/$0.00+$.50

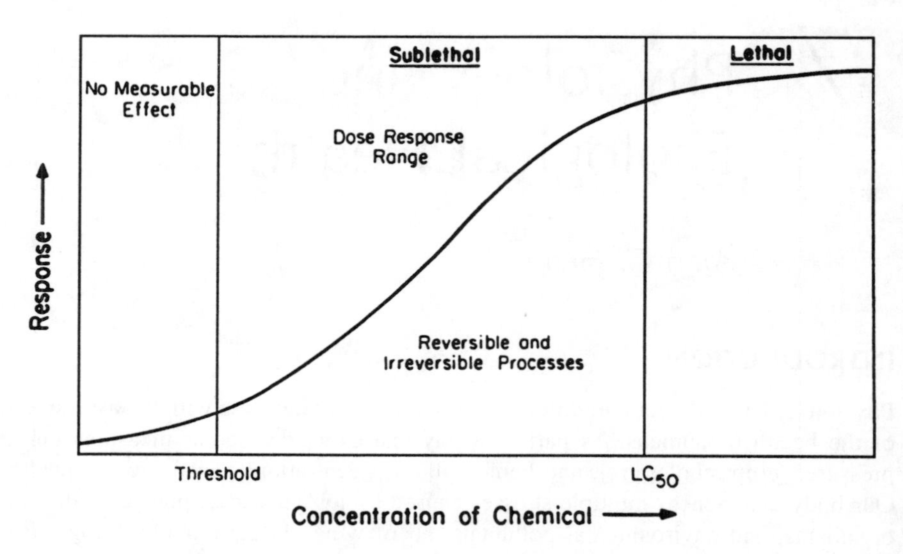

FIGURE 7.1 Idealized diagram of the effect of dose on the response as measured by some physiological change (including percent dead). (Adapted from Waldichuck, M., *Phil. Trans. R. Soc. London,* B, 286, 397, 1979.)

Fish may experience a variety of potential stressors in the environment. These include low dissolved oxygen concentration, lack of food, conspecific social interactions, predation, rapid temperature changes and a diversity of toxic pollutants. The orientation of this book is ecotoxicology so the emphasis here will be on toxic pollutants, however, it must be appreciated that the fish in a natural body of water may feel the impact of one or more of the other factors mentioned as well as the presence of toxic pollutants. Thus, we are really talking about multiple stressors.

THE DOSE–RESPONSE PARADIGM AND ITS RELATION TO HIGHER LEVELS OF BIOLOGICAL ORGANIZATION

When fish or other organisms are exposed to a stressor (e.g., toxic chemical in the water) for a period of time, such as a week, the response depends on the intensity of the stressor. In Figure 7.1 we see that at higher concentrations of the toxic chemical the response is dependent on the intensity of the stress until levels are reached where lethality occurs. Most traditional aquatic toxicology has been devoted to measuring the LC_{50}, the concentration causing lethality to 50% of the test organisms (or as the bumper sticker says "Toxicologists do it 'til they're half dead"). A huge data-base now exists of LC_{50} values for a variety of chemicals and test organisms, and this information is useful for comparing relative chemical toxicities and species sensitivities.

Physiologists who work with toxicity are usually most interested in the part of the curve illustrated in Figure 7.1 referred to as "sublethal," where alterations in

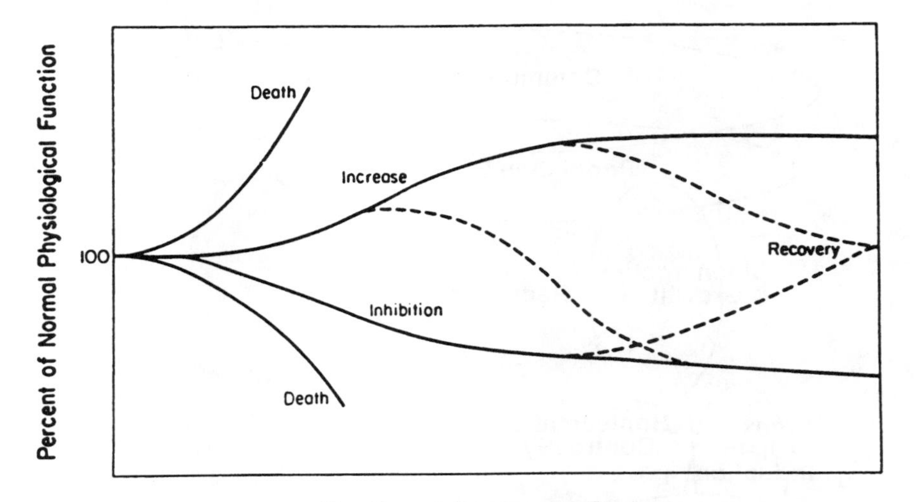

FIGURE 7.2 Generalized illustration of the types of changes that can occur in some physiological variables in fish as a result of exposure to an environmental stressor.

physiology and histological structure are taking place. In order to assess these, the physiologist may measure alterations in blood chemistry (e.g., glucose), organ function (e.g., breathing rate), or histopathology of organs such as gill tissue. Some of these changes may reflect failure of homeostatic mechanisms, or they may be physiological response to the stressor as a mechanism to restore homeostasis.

The character of a physiological response is usually very dependent on how long the organism is exposed to the stressor. Figure 7.2 illustrates some of the types of changes that can be seen. If the degree of stress is sufficiently high, the variable being measured may go one way or the other rapidly, followed by death. This does not necessarily mean that it was the cause of death (or the target of the toxicant), because other variables not measured may be even more important as a cause of death. Mortality from toxic chemicals in fish or other aquatic animals is rarely due to a single mode of action, although there may be a single physiological dysfunction that starts a cascade of other changes.

When exposures are clearly sublethal, the physiological variables being measured in the animal may increase or decrease, usually over a period of hours or days. This may be followed by an establishment of a sort of new equilibrium, or there may be a return to normal, reflecting a form of adaptation to the stressor. At increasing concentrations of toxic chemicals, the general pattern is to move the curves in Figure 7.2 to the left and increase the amplitude of the change (i.e., the dose–response effect).

A toxic chemical in the water or other environmental stressor such as lack of oxygen will first affect molecular functions within cells, altering metabolic activity,

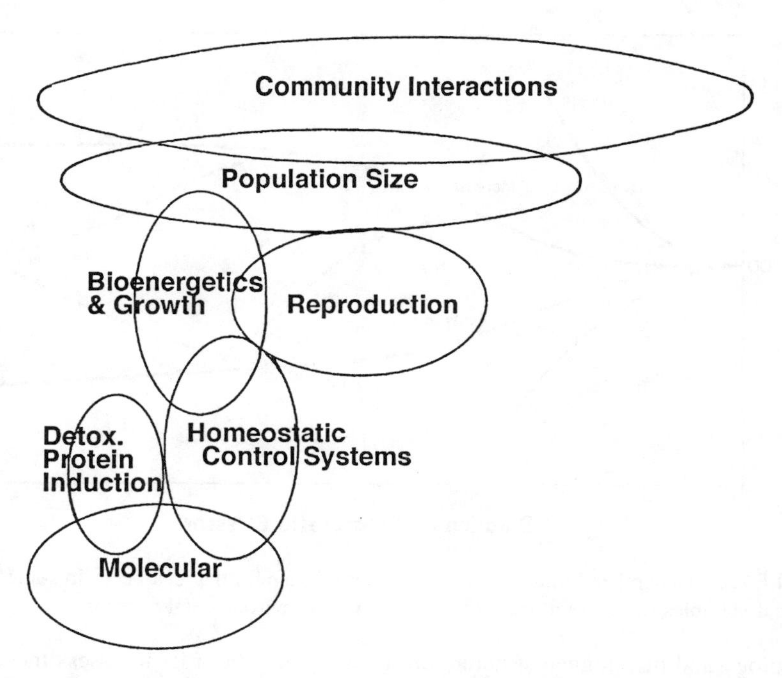

FIGURE 7.3 Interrelationships of changes in different levels of biological organization. Effect of an environmental stressor such as a toxic chemical usually begins at the molecular level and then radiates up through higher levels of biological organization, ultimately impacting the community. See text for further explanation.

permeability etc. If the stress is due to one or more contaminants, detoxification proteins may be induced to sequester or metabolize the toxic chemicals (Figure 7.3). Contaminants or other stressors that affect cellular function will usually then impact homeostatic control systems at the organ and organ system level, and these may or may not be capable of counteracting the stress. As is illustrated in Figure 7.3, these effects on lower levels of biological organization can impact higher levels such as bioenergetics and reproduction. If growth and reproduction become compromised, recruitment into the population may decrease, causing loss of population size. This in turn causes effects on communities such as reduced predation on some aspects of the community and, perhaps, a loss of food for higher predators. Thus, we see that physiological alterations in individual organisms can ultimately have ecosystem effects although the exact connections are often poorly understood (Adams et al. 1992; Heath 1990).

The remainder of this discussion will be devoted to a critical examination of some methods that are or could be utilized in field studies to gain some idea as to how the ecosystem is affecting the residents. My own biases will undoubtedly come out; other researchers may have somewhat different preferences, but perhaps this will generate some discussion. For the purposes of this chapter, I do not distinguish between biochemistry and physiology; rather, biochemistry is used as a tool to measure the physiology of the animals.

TABLE 7.1
Effects of Environmental Stressors on Gills and Associated Structures

Irritation of and damage to lamellar epithelium
Coughing
Decreased oxygen transfer
Increased ventilation frequency**
Decreased chloride cell function and/or increased gill permeability
Altered blood electrolytes and osmoregulation
Decreased blood electrolytes (freshwater)*
Increased blood electrolytes (seawater)*
Gill histopathology
Increased stress protein synthesis**

* Moderately useful for field biomonitoring
** Very useful for field biomonitoring

PHYSIOLOGICAL MEASUREMENTS USEFUL OR POTENTIALLY USEFUL FOR MONITORING ECOSYSTEM HEALTH

Fish physiologists traditionally carry out their measurements on laboratory specimens under carefully controlled conditions, often using very elaborate apparatus. This has yielded tremendous gains in our understanding of the basic physiology of fishes and has been reviewed in three recent books (Evans 1993a, 1997; Rankin and Jensen 1993). Adapting the methodologies of the laboratory physiologist for use on fish in the field, however, often presents challenging difficulties. For example, there is the large problem of capturing fish and obtaining blood and tissue samples before the response to capture and handling masks any effects that the environment had induced. Also, fish do not necessarily stay in the same vicinity, so subjects captured in one place may have just moved in to the area, so their physiology would not reflect the conditions where captured. I will return to these matters toward the end of this discussion.

GILLS

The epithelium of the gills is generally the first organ affected by the presence of a toxic chemical in the water (Table 7.1). Because it has to serve as a respiratory gas exchange organ, the epithelium is extremely delicate compared with the rest of the surface of a fish. Moreover, there is a very large flow of water past the gill lamellae (approximately 70 mL/min in a resting 0.5 kg trout [McKim and Goeden 1982]) so there is ample opportunity for contaminated water to come in contact with this epithelium.

Within minutes, many different waterborne contaminants will cause irritation of the gill epithelium and may stimulate mucus secretion. If contaminant exposure is prolonged or severe enough, swelling may follow (Mallatt 1985), resulting in decreased oxygen transfer from the water into the blood (e.g., Bass and Heath 1977). All these changes generally cause the fish to start coughing and to increase its ventilation frequency.

The technology for measuring ventilatory frequency of fish is well developed (Gruber et al. 1991). Individual fish are held in small aquaria in which electrodes are placed at each end. As the fish breaths, the bioelectric potentials from the ventilatory muscles are detected by the electrodes, and this small (microvolt levels) electrical activity is amplified and then displayed on a suitable recorder. The resulting sine wave can then be used to count frequency. This has now been automated so that a microcomputer can be interfaced with an analog-to-digital converter and used to obtain real-time data of breathing frequency changes. Aberrant breaths such as coughs can also be detected.

While it is currently impractical to monitor ventilation frequency of free-swimming fish in the field, it is quite feasible to divert a small amount of water from a stream into a series of aquaria equipped as described above. Thus, the fish becomes an extremely sensitive sensor for the presence of toxic chemicals in the water, a sensor with a very short time-lag and consequently a first line of defense for water quality monitoring. This system is now available commercially (Biological Monitoring Inc., Blacksburg, VA) and is being used primarily for detecting the presence of toxic chemicals coming into domestic drinking water supplies.

The fish ventilation response to chemical contamination has been used to detect alterations in water quality of trout streams in Tennessee (Morgan et al. 1988). For this purpose, data collection platforms have been constructed to hold a series of rainbow trout ventilation chambers, a water sampler, and pH and temperature sensors. These platforms are placed on stream sides and employ satellite or hardwire transmission of data to a central data coordination and processing facility. Thus, several streams can be monitored simultaneously.

Fish ventilation measurement has considerable potential for further application as an early warning biomonitor. The chief disadvantages are the high cost of the equipment and the fact that it is useful only for detecting fairly rapid water quality changes. The latter problem is due to the fact that fish can often adapt to a low level of stressor and, thus, if the change is very slow, the fish may fail to detect it (Gruber et al. 1991).

Healthy fish must maintain their blood electrolyte concentrations within a fairly narrow range. This category of physiological measurements could be placed later under "blood," but are included here because the chloride cells in the gill epithelium are critical for regulating electrolyte levels in both marine and freshwater fishes (Evans 1993b). When electrolyte dysfunction is observed, it is due to alterations in chloride cell function or altered gill permeability, thus the inclusion here under gill function. In general, when freshwater fish are exposed to xenobiotics they loose blood electrolytes and gain water, whereas marine species show increases in these ions and lose water from their blood (for reviews see Heath 1995; Wendelaar Bonga and Lock 1992).

As part of a general clinical blood chemistry, the plasma ions Na, K, Cl, and Ca have some merit, but for many conditions they tend to show significant changes only with acute environmental stresses. They are, however, extremely sensitive to acid pollution (Wood 1989).

There has only been limited use of changes in blood electrolytes or osmolality in field studies, in part because the relationship between any changes seen in the fish and higher levels of organization is currently obscure (Haux, et al. 1985a; Larsson et al. 1988; Bidwell and Heath 1993). In the future it may be better to use the measurement of blood osmolality alone rather than the individual electrolytes when doing a general blood chemistry. This is because blood osmolality indicates the same general trend as electrolytes and requires only a very small drop of blood for analysis, so it is suitable for work where blood volume is severely limited, as for example with small fish.

The use of stress protein analysis in ecological monitoring is a very new but rapidly expanding field. Stress proteins are a group of cellular proteins that are synthesized in response to virtually any kind of stress. They were previously called heat shock proteins because they were first discovered in *Drosophila* that had experienced a rapid rise in temperature (Welch 1993). They are found in nearly all cells of animals, and their function is to regulate folding, assembly, and aggregation of the other proteins of the cell. They are included here because gill tissue is generally affected before others and might be predictive of subsequent toxic action of contaminants. This is not, however, to suggest that other tissues might not be even more appropriate for study.

Stress protein analysis has thus far been used mostly in crustacea and mollusks (see Sanders 1993 for review). In fish, induction of stress proteins in gills was seen in those exposed to sediments contaminated with a variety of organics and heavy metals (Theodorakis et al. 1992). While this is considered to be a nonspecific biomarker of stress, recent work suggests that the low molecular weight stress proteins have different isoforms, and their induction may differ with the specific stressor (Bradley 1993; Sanders 1993). Thus there is the potential for using this technique to detect fish that are suffering generalized stress and possibly also gain some understanding of the specific environmental cause. Clearly this is an area that deserves further research.

Hematology and Immunology

The term hematology refers to blood plasma chemistry and the study of blood cells. The chemicals that are present in blood are nearly unlimited and many have been analyzed in fish experiencing a variety of stressors (partially reviewed in Hille 1982; Folmar 1993). Only a small number of these chemicals are of value in assessing health of fish in the field, in part because of very large inherent variability (Table 7.2). For example, the plasma enzyme leucine amino naphthylamidase is released from lysosomes in damaged tissues and gets into the blood where it is easily measured. Theoretically, it should be a useful tool for assessing damage from toxic chemicals or other environmental stressors (Bouck 1984). However, Dixon et al. (1985) found that this variable was very sensitive to starvation, reproductive activity, changes in salinity, and the stresses of capture. Perhaps as a consequence of these,

TABLE 7.2
Effects of Environmental Stressors on Hematology and Immunology

Decreased or increased hemoglobin, hematocrit, and erythrocyte number *
Changes in erythron organization**
Altered leukocyte count
Increased plasma cortisol*
Increased plasma glucose**
Immunological depression**
Reduced disease resistance**

* Moderately useful for field biomonitoring
** Potentially more useful for field biomonitoring

the variability between fish specimens receiving the same treatment was large relative to the changes seen with toxicant stress, thus limiting its applicability as a monitoring tool. To a certain extent, the same thing can be said for a variety of other blood chemicals, so a good deal of selectivity by experimenters is needed. For the remainder of this section I will limit the discussion to only certain aspects of blood chemistry; other parts are taken up under different categories (e.g., gills for plasma electrolytes; liver for plasma enzymes).

It is common in clinical studies to measure the differential leukocyte count and the hemoglobin and/or hematocrit (Blaxhall and Daisley 1973). In laboratory investigations, the latter two variables may increase or decrease depending on the type of stressor (for review see Heath 1995; Folmar 1993). Only a very few studies have examined these variables in fish taken from the field where those from polluted waters generally showed a mild anemia (e.g., Casillas et al. 1985; Bidwell and Heath 1993), although fish near a kraft pulp mill exhibited an elevated RBC count (Larsson et al. 1988). There is a great deal of variability in hematocrit values in fish taken from the same place, and the season in which they are collected adds to that variance (Bidwell and Heath 1993). Fish can apparently tolerate a considerable degree of anemia with little apparent harm, so it is difficult to tell whether mild anemia means much biologically (Holeton 1972). For these reasons and the fact that hematocrit or hemoglobin measurement requires a considerable volume of blood (even when using microhematocrit tubes), I do not suggest this measure for assessing fish health in the field, unless rather large fish specimens are available.

Recently, Houston et al. (1993) have developed a more sophisticated way of looking at hematology in fish suspected of experiencing stress. It recognizes that a wide variety of environmental stresses, both natural and anthropogenic, cause increased rates of erythrocyte synthesis. This results in a shift in the population of erythrocytes in favor of immature cells, division of circulating juvenile cells, and karyorrhexis, all of which can have profound influences on respiratory gas transport even though total hematocrit remains relatively unchanged. The technique is based on using blood smears which can be taken in the field and examined later in the laboratory. Another plus is it requires only small quantities of whole blood.

Differential leukocyte counts could also be made on the same blood smear used in the foregoing hematology tests, however, Niimi and Lowe-Jinde (1984) found that long-term (i.e., 75 to 119 days) exposure of trout to methylmercury and chlorobenzenes caused little change in this count. Moreover, they found that fish taken from the field differed markedly from those in the laboratory, which raises the question of what is a normal differential count?

Activation of the hypothalmic–pituitary–interrenal axis results in elevations in plasma cortisol and is a classical way to quantify stress in fish (Donaldson 1981; Thomas 1990). The number of studies where plasma cortisol has been measured (almost all in laboratory or hatchery settings) is considerable (for review see Barton and Iwama 1991; Folmar 1993). The changes in plasma cortisol seen in response to a variety of stressors is large, and the methods for analysis are becoming easier to perform (Barry et al. 1993). However, it is sensitive to the stress of capture so this must be kept constant or reduced.

The argument can be made that modest elevations in plasma cortisol are an adaptation to stress and therefore do not necessarily mean the fish is suffering some dysfunction. Such a conclusion may be justified in many cases, but an elevation in cortisol may also impose an energetic cost on the fish. For example, Davis et al. (1985) found that artificially elevating the plasma cortisol by inclusion of this hormone in the diet of channel catfish was accompanied by a decrease in growth rate even though all fish were fed the same.

The dynamics of the changes in cortisol can be large, as is illustrated in Figure 7.4. As the stress is perceived by the fish, there may be an initial large increase as a sort of alarm response. This may be followed by a return to a lower level, even though the stressor persists. Note in Figure 7.4 that with copper exposure, this initial decrease was then followed by a large increase. This is probably due to the accumulation of copper in the gills, which causes a failure of osmoregulation (for review see Heath 1995) so the delayed rise in cortisol may be an attempt to correct that dysfunction. DDT and atrazine (an organochlorine herbicide) seem to cause no initial alarm in the fish, but as they accumulate, the degree of cortisol response increases.

Measurements of plasma cortisol from feral fish are rare. Giesy (1988) sampled largemouth bass from a reservoir where starvation and thermal stresses were suspected. Cortisol was negatively correlated with body condition, so the results suggest that starvation did not activate this hormone response.

In field work, cortisol probably has potential for assessing acute stress, as for example, determining the degree of stress in fish which are possibly being affected by a chemical spell. On the other hand, the long-term picture may be more difficult to interpret. "Normal" levels of this hormone in fish may indicate they are indeed not experiencing stressful conditions, or, the fish may have adapted to the presence of contaminants. But as these contaminants accumulate in the body of the fish over weeks or months of chronic exposure, internal harm might occur without changes in this so-called stress hormone, so its value as a predictor of environmental impacts on population size may be limited.

Chronic exposures to some contaminants may even compromise normal cortisol response to stress. In a recent study (Hontela et al. 1992) fish were taken from sites

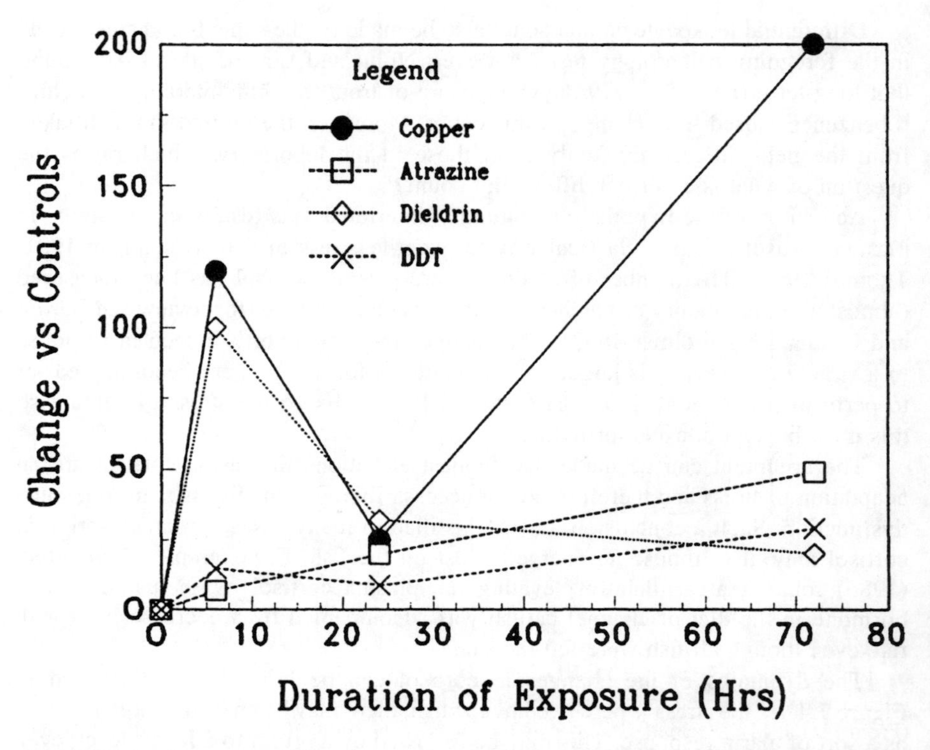

FIGURE 7.4 Some examples of alterations in plasma cortisol concentration relative to unstressed controls. (Data from Guth, G. and Hanke, W., *Ecotox. Environ. Safety*, 9, 179, 1985, and Schreck, C. B. and Lorz, H. W., *J. Fish. Res. Bd. Can.*, 35, 1124, 1978.

in a river that were highly polluted with polycyclic aromatic hydrocarbons, polychlorinated biphenyls, and mercury. The specimens from the polluted sites were unable to show an elevated cortisol in response to the stress of capture, and their pituitary corticotropes were atrophied. The authors suggest the fish had exhausted their cortisol-producing endocrine system, perhaps as a result of prolonged hyperactivity by the corticotrope or interrenal cells.

Another traditional indicator of acute stress is an elevation of blood glucose from mobilization of liver glycogen (Barton and Iwama 1991). It has long been assumed that cortisol is a key hormone stimulating this process because the two variables are often correlated (Van der Boon et al. 1991). However, recent findings have cast some doubt on that assumption. Anderson et al. (1991) fitted rainbow trout with osmotic pumps which maintained an elevated plasma cortisol (two- to fourfold above the controls) for 10 d. This treatment caused no effect on plasma glucose nor on hepatic glucogenic enzymes, so the hyperglycemia seen with stress in this species may be caused by catecholamines rather than cortisol.

Although the direct cause of hyperglycemia may be in question, there seems little reason not to measure it in fish sampled in the field if acute stress is suspected because the amount of blood plasma required is small and the procedure is simple.

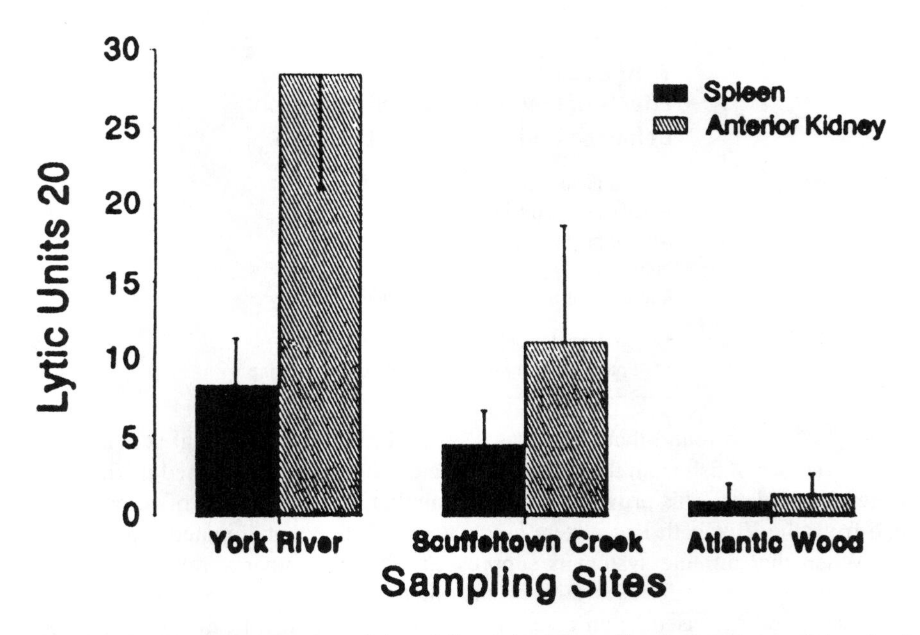

FIGURE 7.5 Natural cytotoxic activity of *Fundulus heteroclitus* leukocytes as determined by 4 h ^{51}Cr-release assay at 100, 50, and 25 effectors: 1 K562 cell. The percent of specific release was transformed into lytic units. Data are expressed as means ± standard deviations. (From Faisal, M., Weeks, B. A., Vogelbein, W. K., and Huggett, R. J., *Veterinary Immunol. Immunopath.*, 29, 339, 1991. With permission.)

Because of this, it may at times be preferable to measure this instead of cortisol because the latter is much more difficult and expensive to analyze, although the percent changes seen with a given degree of stress are usually greater with cortisol.

Some investigators might maintain that catecholamines should be measured. However, their analysis is currently difficult and expensive. Moreover, they are extremely sensitive to the effects of capture and handling.

Vertebrate animals under stress often exhibit a suppressed immune system. The mechanism of this undoubtedly is at least in part due to elevations in cortisol. Several studies have shown that cortisol at levels only slightly higher than those found in unstressed fish can suppress various immune functions (Thomas and Lewis 1987; Pickering and Pottinger 1989). Also, toxic chemicals can affect immune activities directly (Anderson 1990; Weeks et al. 1992). For field work, one of the advantages of using immune function for assessing fish health is that any changes associated with capture and handling are delayed, so if blood or tissues are obtained promptly, that is not an important factor.

An example of the use of immune function to assess condition of fish from a polluted river system is presented in Figure 7.5. *Fundulus heteroclitus* were sampled from two sites in the Elizabeth River of Virginia which are contaminated with polynuclear aromatic hydrocarbons in the sediments. For comparison, others of the same species were taken from a reference site on the York River, Virginia. Faisal

TABLE 7.3
Effects of Environmental Stressors on Behavior and the Nervous System

Altered locomotor activity*
Avoidance of pollutants*
Altered predator–prey interactions
Inhibition of olfactory sense
Inhibition of brain acetylcholinesterase activity**

* Moderately useful for field biomonitoring
** Potentially very useful for field biomonitoring

et al. (1991) also found that there was a gradual recovery of normal activity of the leukocytes when fish from the two contaminated sites were maintained in York River water for 28 days. This provided confirmation that the lower levels of activity in the fish from the Elizabeth River were associated with the contaminated sediments.

When the immune system is suppressed, fish (and other vertebrates) become more susceptible to diseases they might otherwise be able to overcome. Disease surveys have been used as an assessment technique for fish health in several ecosystems (Sindermann 1979; Goede and Barton 1990). O'Connor et al. (1987) developed an index of disease conditions (primarily external) that may be induced in marine teleosts and shellfish from pollution. They carry their work to a further step that is important but rarely done. They attempt to provide information so that nonprofessionals can interpret the biological significance of the changes seen in the fish.

The potential for the use of immune system and disease measures is probably large, especially as newer methods are developed that do not require so much expertise or equipment (Weeks et al. 1992).

Behavior and Nervous System Function

Behaviors are basically controlled by the nervous system, but they are influenced by a multitude of factors both around and within the animal. The animal is constantly sampling the environment through the sensory receptors, and this information is processed by the nervous system and therefore influences behaviors (Table 7.3). Within the animal, changes in blood chemistry and hormone levels that are caused by multiple stressors in the environment also affect the function of the nervous system. Thus, the nervous system and the behaviors that it produces can be viewed as an integrator of multiple effects on the animal.

Behavioral toxicology has received a considerable amount of interest in recent years (for reviews see Beitinger 1990; Blaxter and Hallers-Tjabbes 1992; Little et al. 1993). Random locomotor activity in fish is quite sensitive to a variety of pollutants and has been quantified using techniques that range from simple to extremely sophisticated (reviewed in Heath 1995; see also Little et al. 1990). Quantifying locomotor activity in the field would be quite difficult, but fish could be tested in the laboratory using water from the field, or conversely, feral fish might be brought into the lab for testing. This would be quite feasible as part of a larger program in which other variables were measured. An advantage of behavioral measures is that they are noninvasive, so species that may be in limited quantity could be tested and then returned to the field.

Fish have the ability to detect and then possibly avoid many pollutants, sometimes at far lower concentrations than those that cause mortality. This ability has been quantified for a wide variety of pollutants using laboratory devices that range from simple to complex (for a review see Heath 1995; Beitinger 1990). Avoidance has also been confirmed in a few field studies. For example, Gray (1990) used biotelemetry of salmon to document avoidance of oil-contaminated water. Clearly, if fish are avoiding an area, that is a pretty good indication that the area is unhealthy for fish. Because of this, it is now legally accepted as evidence of injury for Natural Resource Damage Assessment under proceedings of the Comprehensive Environmental Response, Compensation and Liability Act of 1980 (NRDA 1986). It must be recognized, however, that avoidance behavior may not protect against all contaminants. For example, dissolved mercury is apparently not avoided and may even act as a mild attractant for fish (Black and Birge 1980).

The enzyme acetylcholinesterase hydrolyzes the neurotransmitter acetylcholine so that it does not build up in concentration around synapses where it is produced. This enzyme is sensitive to the presence of organophosphorous and carbamate pesticides, and its activity is easily measured. Thus, suppression of activity is a useful biomarker for this sort of poisoning (Mineau 1991). However, other pollutants such as mercury have also been found to cause some inhibition of acetylcholinesterase in the brain of contaminated fish (Shaw and Panigraphi (1990). It appears that 70 to 80% inhibition of activity must take place before death of a fish (Coppage and Mathews 1974; Gantverg and Perevoznikov 1984). But less extensive inhibition is associated with behavioral abnormalities such as hyperactivity and loss of equilibrium which might prove lethal in an ecological setting as opposed to a laboratory aquarium (Zinkl et al. 1991). Thus, while measurement of acetylcholinesterase activity has no value as an indicator of general health, it could be useful in a diagnostic sense to narrow down the possible sources of environmental stress.

LIVER

Of all the organs in the fish's body, the liver and gills probably experience the greatest degree of harm from chemical stressors. The liver is especially sensitive to contaminants because these chemicals tend to accumulate there to much higher concentrations than in other organs, so the liver cells experience toxicity often before those in other organs (Table 7.4). In addition, the liver serves numerous functions associated with energetics and the metabolism of harmful chemicals. Thus, alterations in liver function can reflect both dysfunctions and adaptations to stressors. Many of the measures of liver function are insensitive to the stress of capture and handling so they lend themselves to field surveys.

The liver size is greatly affected by nutritional level and the presence of organic toxic chemicals. The liver/somatic index (often called the hepatosomatic index, HSI) is the percentage that the wet weight of the liver makes up of the whole body mass.

TABLE 7.4
Effects of Environmental Stressors on Liver

Increased or decreased liver–somatic index**
Decreased lysosomal membrane stability**
Increased plasma enzymes from liver*
Decreased tissue ascorbate concentration**
MFO enzyme induction (e.g., EROD)**
Biotransformation of toxic chemicals
Metallothionein induction*
Metal sequestering and excretion
Altered tissue glutathione concentration*

* Moderately useful for field biomonitoring
** Potentially very useful for field biomonitoring

It is used widely as a measure of general condition in fish captured in the field (Goede and Barton 1990). Decreases in HSI are frequently seen in fish under chronic stress and are often correlated with loss of energy stores (Lee et al. 1983; Goede and Barton 1990).

Increases in liver mass may be seen in fish that have been exposed for long periods of time to organic contaminants such as petroleum hydrocarbons (Sloof et al. 1983; Vignier et al. 1992; Everhaarts et al. 1993). It is usually associated with increased capacity to metabolize organic xenobiotics, so is actually an adaptation to pollution stress. Obviously, the nutritional and organic pollutant effects are in opposition as regards HSI, so some interpretation is required when obtaining these sorts of data. For example, if the fish are experiencing organic pollutant contamination but food is limited, a decrease in HSI may occur even in the presence of a stimulant for hypertrophy.

Lysosomes are membrane-bound cellular organelles that contain acid hydrolases. The stability of the lysosome membrane exhibits nonspecific sensitivity to environmental stressors, so finfish or shellfish that are under stress have lysosomes with abnormally low stability (Mayer et al. 1992). This can be quantified using a biochemical procedure (Versteeg and Giesy 1985). Kohler (1991) assessed pollution in contaminated estuaries with this technique and found decreased lysosomal stability in a gradient that correlated with the level of pollution. Since lysosomal membrane stability is affected by both chemical and nonchemical stressors, it serves as a sort of integrator of multiple stressors so its potential for field work is probably high.

When liver cells start to break down due to toxic chemical contaminants, they release some of their enzymes into the bloodstream. Clinical laboratories have long used the measurement of the activity of these enzymes in the blood plasma (sometimes called serum) as a diagnostic tool for assessing liver damage in humans and domestic animals. The enzymes of primary interest are glutamic oxaloacetic transaminase (GOT), glutamic pyruvic transaminase (GPT), and sorbitol dehydrogenase (SDH). These are normally of low concentration in the plasma. High plasma SDH activity is specific for liver damage, whereas the others are also released by other damaged tissues (Dixon et al. 1987). Assays of these enzymes may ultimately have utility for evaluation of fish in the field, but considerably more work is needed to understand the effects of multiple exposures to compounds, how temperature and diet affect them, and the long-term consequences of altered levels of these plasma enzymes (Mayer et al. 1992).

Ascorbate is probably better known as vitamin C. It is involved in numerous functions in cells such as collagen synthesis, prevention of the buildup of free radicals, and metabolism of steroids. It is included here in the section on liver because it has been measured mostly in that organ of fish. However, it is found in nearly all tissues. Different fish species may or may not have a requirement for ascorbate in the diet. Salmonids, channel catfish, and mullet require it in the food, but cypriniformes, which include carp, do not (Thomas et al. 1982; Halver et al. 1975).

Some laboratory experiments have demonstrated that supplementary amounts of ascorbate in the diet reduces the effects of toxic metals (Yamamato et al. 1981) and some pesticides (Verma et al. 1982). Chronic exposure to some pollutants has been shown to cause sizable declines in liver ascorbate concentration. Six-month exposure to carbofuran, a carbamate insecticide, produced a 44% decline in Indian catfish (Ram and Singh 1988), and a one-week exposure to No. 2 fuel oil caused a 45% reduction in mullet (Thomas and Neff 1984). In the latter study, Thomas and Neff also showed that the stress of capture has little effect on this variable. However, Bidwell and Heath (1993) measured ascorbate in freshwater rock bass (a centrarchid) *in situ* at various times of the year and found a considerable seasonal and sexual effect.

Chronic exposures of fish (and most other animals) to various pollutants may cause the induction of three classes of cellular proteins: stress proteins, biotransformation enzymes, and metallothioneins. The so-called stress proteins, already discussed in the section on gills, serve primarily for prevention of damage and repair of proteins. Here we take up the process whereby the animal attempts to biotransform organic xenobiotics and to sequester metals. The induction of stress proteins, biotransformation enzymes, and/or metallothioneins represents biochemical adaptation to the presence of specific pollutants. One could argue that detection of these processes alone does not necessarily indicate fish that are unhealthy or under stress, but it can be an early warning as it certainly suggests contamination. Furthermore, these adaptations have a limit that, if exceeded, leaves the animal vulnerable to toxic stress. They could also impose an energetic demand that may lower the energy available for growth.

Metabolic biotransformation of organic xenobiotics usually involves a phase I process that is commonly, but not always, an oxidation catalyzed either by flavoprotein monooxygenases or cytochrome P450. The product of the phase I metabolic activity is then conjugated in phase II reactions to an endogenous compound such as glutathione, although sometimes phase I steps are bypassed with the xenobiotic going directly to the phase II stage. The ultimate product of these metabolic transformations is generally less toxic (but sometimes more carcinogenic) and more water soluble and thus more readily excreted. The biochemistry and molecular biology of the processes of biotransformation have received a huge amount of attention in the past 15 years and have also been frequently reviewed (e.g., Payne et al. 1987; Melancon et al. 1988; Stegeman and Hahn 1994).

Induction of the biotransformation enzymes is most frequently measured in liver by determining the activity of certain enzymes such as ethoxyresorufin-O-deethylase (EROD) or the amount of cytochrome P450. There are actually several forms of cytochrome P450 in fish, some of which are induced by specific chemicals, so they can be used in a diagnostic sense (Stegeman and Hahn 1994). However, their analysis is more difficult than measuring a more general indicator such as the activity of EROD.

The induction process is fairly rapid as is seen in Figure 7.6. In this study, carp were exposed to water pumped directly from the Kinnikinnic River, Wisconsin, and control carp received dechlorinated tap water. By three days of exposure to the river water containing PCBs, there was a fivefold increase in EROD activity in the liver, and this reached a peak sometime after 10 days. Note that some of the exposed fish were placed in noncontaminated water after 22 d, and the EROD activity declined, reaching the control level sometime between 8 and 25 d. later. This return of EROD activity to the control level is not always seen. In the same study from which Figure 7.6 was taken, the researchers captured feral carp from a highly contaminated river (pollutants not given) and kept them in the laboratory. Even after 98 d in the clean laboratory water they still exhibited a greatly elevated EROD activity. Investigators attribute this to possible genetic differences. However, it seems to me that the feral fish might also be contaminated with compounds in the fat of the fish that depurate extremely slowly, so these could continue to act as an inducer.

Measurement of monooxygenase activity has been used fairly extensively in field biomonitoring of pollutant impacts on marine and freshwater fishes (reviewed in Payne et al. 1987). One of the virtues of this technique is that the levels of activity are not affected by the stress of capture so it lends itself to work on feral fish. Furthermore, it is extremely sensitive to organic pollutants, perhaps more so than any other single measurement. Monooxygenase activity is even inducible in eggs and fry, and this has been seen in both field and laboratory work (Binder and Lech 1984).

Metallothioneins (MTs) are low-molecular-weight intracellular polypeptides (sometimes referred to as proteins) that have a strong affinity for metals. They are found in both plants and animals; at least 24 species of teleosts and elasmobranchs have been reported to have them (Roesijadi 1992). In fish, MTs have been found in intestine, gills, kidney, and liver with the latter receiving the major attention. MT binds both essential metals (e.g., iron) and toxic metals such as mercury. Since it is usually in the saturated state, in order to sequester additional toxic metal, more MT must be synthesized (Hamer 1986). It is this induction of additional MT in fish that can be used as a monitor of metal contamination (Roesijadi 1992).

The use of MT induction in monitoring programs has some problems associated with it. Probably the major one is that the concentration of MT is influenced by environmental and seasonal factors independent of exposure to metals (Engel and Brouwer 1989). Thus, the use of appropriate reference animals is critical. Also, from the standpoint of this chapter in which monitoring of multiple stressors is emphasized, this would have a lower priority than some other measures which indicate a more generalized presence of stress, rather than the presence of metals alone.

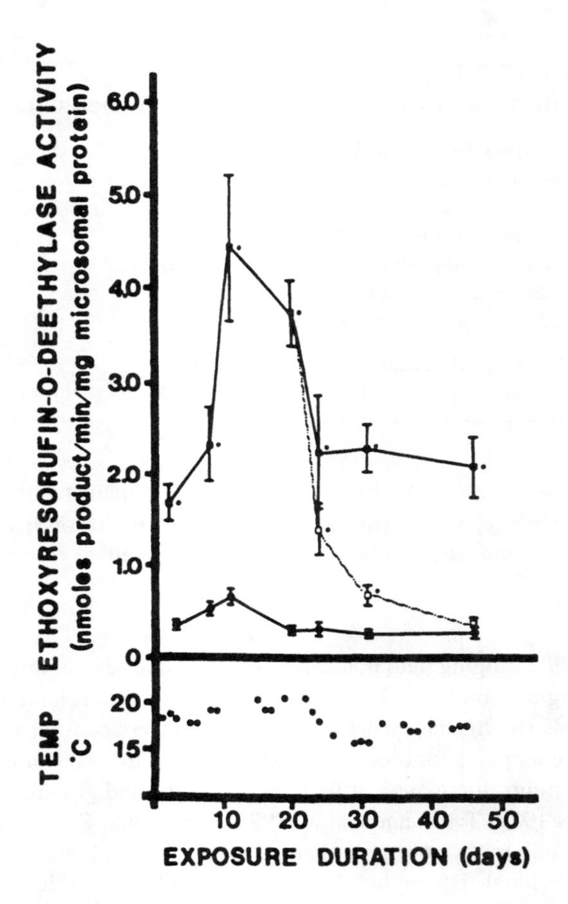

FIGURE 7.6 Effect of exposure to river water on hepatic microsomal monooxygenase activity in carp. Hepatic microsomes were prepared and assayed for ethoxyresorufin-O-deethylase activity. Solid circles = carp maintained in dechlorinated city water. Solid squares = carp exposed to river water. Open squares = carp exposed to river water for 22 d, after which they were transferred to dechlorinated city water. Vertical bars indicate SE; asterisks indicate significant difference from values for carp maintained in dechlorinated city water. The lower part of the figure shows the average ambient temperature for the carp during the experiment. (From Melancon, M. J., Yeo, S. E., and Lech, J. J., *Environ. Toxicol. Chem.,* 6, 127, 1987. With permission.)

Another cellular peptide is glutathione, which performs a variety of critical functions including the metabolism of peroxides and free radicals. It also binds to metals and plays a roll in phase II detoxification reactions of organics. The overall concentration of glutathione generally rises in the presence of metals and organics (Thomas and Wofford 1984), and the changes can be quite large (Thomas and Juedes 1992). The ratio of reduced glutathione (GSH) to its oxidized form (GSSH) can also change due to the scavenging of radicals or increased peroxidase activity (Stegeman et al. 1992). The relationship of changes in total glutathione or the forms of this

TABLE 7.5
Effects of Environmental Stressors on Energetics

Decreased feeding and growth
Decreased cellular enzyme activities
(e.g., cytochrome oxidase, citrate synthetase)**
Decreased RNA or RNA/DNA ratio**
Decreased energy stores (lipid, protein, carbohydrate)**
Decreased adenylate energy charge*
Reduced swimming ability and stamina**

* Moderately useful for field biomonitoring
** Potentially very useful for field biomonitoring

tripeptide to higher levels of biological organization remain to be elucidated, but since it is relatively easy to measure in liver tissue, its inclusion may be warranted in monitoring programs where contaminants are suspected.

ENERGETICS

Energetics is a wide-ranging field that can be examined at several levels of biological organization, ranging from the cellular to the ecosystem. Our orientation here is how individual animals (in this case fish) can be used to assess ecosystem health. At the individual level, energetics (or bioenergetics) deals with the rates of acquisition (i.e., feeding), partitioning, storage and utilization of energy, and growth (for reviews see Brett and Groves 1979; Tytler and Calow 1985; Adams and Breck 1990). Table 7.5 lists some of the major physiological changes associated with energetics that can be measured on individual fish captured in the field. Other variables, such as rate of energy utilization (measured as oxygen consumption), are usually measured only in the laboratory. Energetics is very much involved with growth; indeed, from a practical standpoint, it is growth that is the main variable of concern in most fisheries studies. Multiple stressors can depress growth of individuals which can then affect recruitment into the population and therefore the community structure (Fausch et al. 1990).

A characteristic of most fish that are under stress is that they decrease their feeding rate (Heath 1995). Of course, reduced food availability in the habitat can also reduce feeding rate. The estimation of food consumption using gut analyses has been extensively discussed from both a theoretical and practical standpoint by Adams and Breck (1990). The methodology is not as straightforward as it sounds because there are several factors that must be considered in setting up a monitoring scheme using this variable. For example, the rate of digestion must be factored in, and this differs with various foods, species, and temperature. Variability from day to day of fish in the same population can also be considerable.

The liver somatic index has been shown to correlate with growth changes that are induced by changes in seasonal temperature or food availability (Adams and McLean 1985). However, as was mentioned above in the section on the liver, this index is also sensitive to the presence of some organic contaminants which can, on occasion cause an increase in liver size in fish under stress. This phenomenon must be kept in mind when using this variable where multiple stressors which include toxic chemicals may be present.

Recently there has been an increasing interest in the use of biochemical measures to estimate feeding and growth rates in fish. These measurements are based on the thesis that protein synthesis, growth, and oxidative metabolism are linked (Houlihan et al. 1993). The activities of the oxidative enzymes citrate synthase and cytochrome oxidase and the glycolytic enzyme lactate dehydrogenase track growth rate closely (Mathers et al. 1992). Given this correlation between enzyme activity and growth in laboratory specimens, samples can then be taken from feral members of the same species, and these data can be used for estimation of growth in the fish from the field.

The recent growth of fish has also been assessed by measuring the RNA (or RNA/DNA) concentrations in tissues, especially white muscle (Busacker et al. 1990). This is especially useful for assessing growth in larval and juvenile fish (Bulow 1987). Limited information suggests that it requires approximately two weeks for RNA and oxidative enzyme activities to stabilize at a new level after the animals have been feeding following a period of starvation (Houlihan et al. 1993), so the biochemical measurements would seem to be relatively immune to the daily changes of food consumption that may occur in fish in an environment where food availability is spotty. More rapid changes in RNA have been found in larvae exposed to toxic chemicals. Barron and Adelman (1984) observed a significant drop after only 4 days of exposure to sublethal concentrations of several different organic and inorganic toxicants and, more important, this was predictive of the impact of the metal on 28 days of growth. The analysis of nucleic acids in individual larval fish is now moderately easy to do, so over 50 samples can be analyzed in a day (Heath et al. 1993). Finally, whole-body trypsin activity (a proteolytic digestive enzyme) has been assayed in larvae and found to correlate well with feeding status and growth potential (Ueberschar 1988). In summary, it is clear that a variety of biochemical techniques are now available to assess the feeding and recent growth rate of feral fish.

The three energy stores of fish are affected by food availability and by the presence of toxic chemicals. Their analysis is fairly easy (compared with things such as MFO enzymes), and protein and lipid concentrations are not affected greatly by the stress of capture. Blood glucose and liver glycogen, however, are sensitive to handling of the fish, thus limiting their use unless caged fish are being sampled.

As a group, fish use protein for energy more than do other vertebrates, so it is fairly sensitive to multiple environmental stressors that may impose an energy drain on the animals (Brett and Groves 1979). Lipids, the other member of the triad, are, of course, an important source of chemical energy, and their concentration reflects the physiological capacity of the fish. The analysis of these energy stores is reviewed in Busacker et al. (1990).

There has been some interest in using the concentration of adenylates to assess the energy status of aquatic animals and microorganisms. The term adenylates refers to the high-energy nucleotides adenosine triphosphate (ATP), adenosine diphosphate (ADP), and adenosine monophosphate (AMP). The adenylate energy charge (AEC) is defined as: (ATP + 1/2 ADP)/(ATP + ADP + AMP). The AEC can vary from 0 to

1; values >0.8 indicate optimal conditions. Tissues from invertebrate animals with AEC from 0.5 to 0.7 are considered to indicate nonoptimal conditions, and <0.5 indicates severe stress (Livingston 1985). The analysis of adenylates from molluskan tissues in field studies probably has some promise (Livingston 1985; Mayer et al. 1992). However, fish AEC values differ considerably with the specific tissue (Heath 1984). Liver generally has a considerably lower AEC than does muscle, yielding values between 0.7 and 0.8 in controls. Muscle ATP concentration is also well regulated, so the environmental stress must be large for it to be affected, but, handling the fish can within seconds cause changes in this variable. Unless one is working with caged fish, the measurement of adenylates from fish tissues probably would have little meaning in a monitoring program. Further discussion of the pros and cons of this measurement can be found in Mayer et al. (1992).

Swimming capacity is important for fish in maintaining position in a current, escape from predators, and for carnivorous fish, in capturing prey. A variety of water tunnels and other arrangements have been developed to quantify different aspects of swimming capacity. Essentially, there are two types of capacity that may be important: stamina (i.e., ability to maintain high speed) and maximum speed. The latter is the more usual one measured because it is easier to be consistent between tests; it also has more relevance for things like evading predation.

From the standpoint of this chapter, swimming performance can be used to assess whether chronic exposures to toxic substances in a body of water impair the capacity of the fish to function in their environment (Heath 1995). In addition, specific pollutants can be tested in the laboratory as to their effects on this function. Pollutants that impair gas exchange in the gills will reduce the aerobic capacity and thus maximum sustained swimming speed. Aluminum is an example of one such pollutant. (Wilson and Wood 1992). Other pollutants that are neurotoxic can affect swimming coordination (see Peterson 1974). However, depending on the type of swimming capacity measured, pesticides may or may not affect swimming (Little et al. 1990).

Only limited use of swimming capacity measurements has been made in assessing the condition of fish from natural waters. Striped bass from the Hudson River, New York, were compared with those from hatcheries and other noncontaminated east coast rivers. Those from the Hudson River exhibited a lower maximum swimming speed, which was presumably due to the accumulation of PCBs in their bodies and to the presence of parasites in their livers (Freadman et al. 1985; Buckley et al. 1985). The technology required to do these kinds of tests can be considerable, although for larvae, very simple systems involving only a Petri dish and stopwatch are feasible (Heath et al. 1993).

Some Practical Considerations in the Use of Physiological Measurements

There are two biological issues of prime importance when attempting to utilize physiological measurements for assessing ecosystem health. First, fish are very sensitive to the process of being captured and handled. This causes a variety of stress responses that can easily mask many variables of interest, mostly those associated with blood chemistry (Pickering et al. 1982; Haux et al. 1985b). Throughout this discussion, I have pointed out which measurements are sensitive to the stress of capture. If there is a need to use these measures, probably the best way to minimize this problem is to use caged fish. Fish can be captured with electroshocking or netting and then placed back in the water body from which they were taken but confined in cages. After about a week they will have recovered from the stress of capture and can then be sampled much like those in an aquarium (Pickering et al. 1982; Haux et al. 1985b). When working with some species that do not take to confinement with conspecifics, modified minnow traps work well as cages for holding individual specimens (Bidwell and Heath 1993). This also helps reduce the likelihood of vandalism because these small cages are more easily concealed than a single large one.

The second major biological issue that must be confronted is that fish are mobile organisms and do not necessarily stay in one area. For some species, the home range is small; for others, large. For those with a large range, it is not possible to say how long they have been in the area from which they are captured. Thus, some knowledge of the behavioral ecology of the species in question is necessary. For those with a limited home range, the problem is minor. But if it is desired to work with a species with a larger or unknown home range, then the use of caged specimens from hatcheries or pristine environments may be the best approach (Grizzle et al. 1988; Haasch et al. 1993).

It is difficult to define normal ranges or baselines for various physiological parameters in hatchery stocks of fish that are fairly uniform genetically (Miller et al. 1983). The problem becomes nearly impossible for feral specimens, so one should always use one or more control or reference stations for comparison. Seasonal changes will be superimposed on other variables in the environment such as anthropogenic contamination, floods, etc. (Bidwell and Heath 1993), so the selection of suitable reference sites is critical.

TWO CASE STUDIES

It would be nice if we could find a single physiological measure that would indicate the health of the fish in a body of water. The search for such a single measure is like that conducted for the Holy Grail and will probably be as fruitless as that was. Thus, there is a need for selecting a wide variety of assessment tools. It is also desirable to have tools at different levels of biological organization because we greatly need to understand the connections between these levels.

A stream in Tennessee was extensively investigated using some of the techniques discussed in this chapter (Adams et al. 1992). The stream is a third- through fifth-order stream that receives a single point discharge of several contaminants such as PCBs and mercury. Four stations were established, C1 being immediately below the industrial outfall, the other three stations ranging downstream 17km. There were decreasing downstream concentrations of mercury and PCB in the tissues of sunfish collected from the four stations. A reference stream was chosen in another watershed that was physically similar but not contaminated. Redbreast sunfish (*Lepomis auritus)* were collected, and a variety of measurements taken on them, all variables

TABLE 7.6
Statistical Comparisons of Physiological Parameters at Each Contaminated Stream Site Compared to the Reference Stream

	Sample Sites			
Physiological Response	**C1**	**C2**	**C3**	**C4**
EROD	+	+	+	+
Serum triglycerides	–	–	–	0
Liver parasites	+	+	0	0
Macrophage aggregates	+	+	+	0
Liver–somatic index	+	+	0	0
RNA/DNA	–	–	–	0
Fecundity	–	0	0	0

Data from Adams, S. M., Crumby, W. D., Greeley, M., Ryon, M. G., and Schilling, E. M., *Environ. Toxicol. Chem.*, 11, 1549, 1992.

unaffected by capture. This species fortunately has a small home range so cages were not needed.

Table 7.6 shows some of the variables and how they compare with the reference stream. Clearly there is some organic contamination at all stations as indicated by the elevated EROD activity, a molecular measurement. The reduced serum triglyceride levels suggests that those fish from the first three stations have lowered energy available and this may help explain the reduced growth rate reflected in the lower RNA/DNA ratios in these same stations. Damaged tissue was also seen in the first three stations as indicated by macrophage aggregates. Fecundity was only affected at station C1, but other data (not shown here) indicated growth impairment in stations C1 through C3. Overall, it appears that by station C4 the stream has undergone nearly a complete recovery so the fish health there is nearly normal. This study shows how physiological variables can be used to define pollution gradients in a stream and especially to evaluate how they are affecting one of the resident organisms.

In the second case study discussed here, a different suite of variables was used to evaluate the impact of contaminated sediments in sites in Puget Sound, Washington (Stein et al. 1992). Two of the sites are urban (Duwarmish Waterway and Hylebos Waterway), with sediments contaminated with a variety of organic and inorganic compounds. Everett Harbor is also urban but less contaminated. Polnell Point and Pilot Point are relatively uncontaminated locations. Juvenile fish were collected because they have small home ranges. The variables measured were PCB concentration in liver, fluorescent aromatic compounds in bile, hepatic aryl hydrocarbon hydroxylase and EROD activity, liver glutathione, and DNA adducts in liver. For each fish species, data for a given variable were normalized to give a maximum response of 100. For each composite the normalized responses were then summed to yield a cumulative bioindicator response. The data are summarized in Figure 7.7

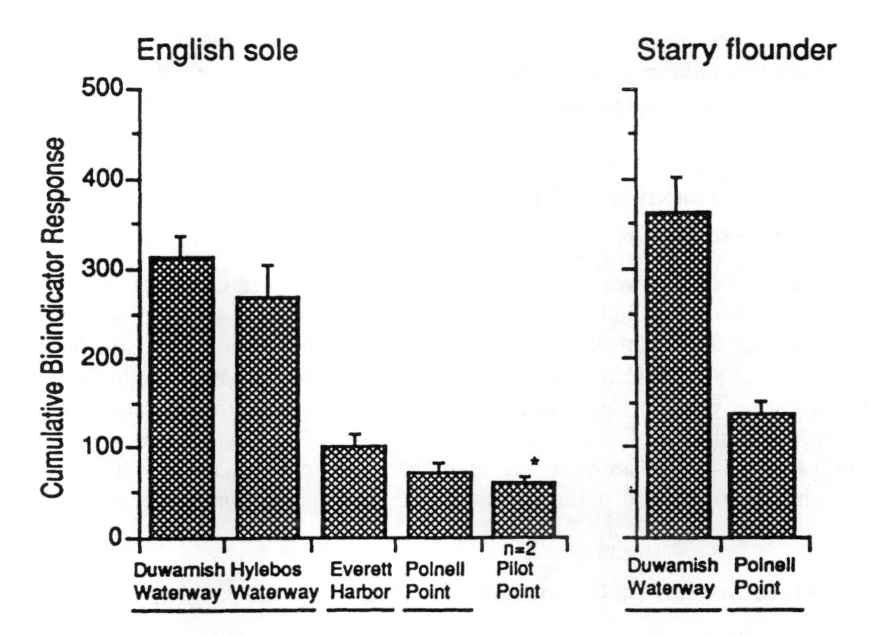

FIGURE 7.7 Cumulative bioindicator response (CBR) as mean ± SE (N = 4 unless otherwise indicated) for English sole and starry flounder sampled from urban (DW, HW, and EH) and relatively uncontaminated (PiPt and PoPt) sites in Puget Sound, Washington. For each fish composite, the CBR is the sum of the normalized response for each bioindicator. Common underline indicates means that were not significantly different (p>0.05). The * indicates means (n = 2) not included in the statistical analyses that were presented for comparison. (From Stein, J. E., Collier, T. K., Reichert, W. L., Casillas, E., Hom, T., and Varanasi, U., *Environ. Toxicol. Chem.,* 11, 701, 1992. With permission.)

where a clear-cut difference is seen between the contaminated and noncontaminated sites. The argument might be made that the same information could be obtained by doing chemical analyses of the sediments. However, this would not indicate how bioavailable the materials were and whether they were doing any harm to the fish. This study does have the weakness of most of its type in that there is no connection shown between the alterations seen in the individual fishes and population level effects. That is a major step that must ultimately be taken but will be extremely difficult.

PHYSIOLOGY CAN BE USEFUL FOR:

Detecting Stress in Fish or Large Invertebrates Before the Population Crashes

When an ecosystem is receiving multiple contaminants, it is critical that we detect their effects before the populations of resident organisms experience severe declines which may require several generations to become evident. Also, waiting until there are dead fish floating is completely unacceptable. When that happens we have clearly failed. As has been outlined in this chapter, there are many sublethal changes in the organisms that indicate stress before the population numbers are reduced, and these can be used to predict subsequent, more permanent harm so corrective action can be taken.

Detecting Gradients and "Hot Spots" of Pollution in Streams and Estuaries

If there is only one known toxic pollutant in a body of water, chemical analyses can evaluate gradients and hot spots. However, more often than not there are multiple pollutants and some of these are not known ahead of time. Moreover, simple chemical analyses do not reveal to what extent the chemical is bioavailable to the organisms from the water or sediment, nor do they tell whether the chemicals are at a concentration sufficient to do biological harm. As was shown in the Adams et al. (1992) study discussed above, the resident fish in a stream can serve as an integrator of the impact of several pollutants coming from a point source or from nonpoint sources.

Evaluating Success of Corrective Measures

When the pollutant input to a body of water is reduced, it is important to know if the organisms are benefiting from the corrective measures. Waiting until the size of the populations changes can take several generations, and stochastic processes may prevent us from ever knowing whether we have done any good. By doing before and after physiological measurements, important species of fish and invertebrates can help reveal the extent to which the ecosystem has been altered.

Evaluating Whether Aquatic Organisms Are Adapting to a Chronic Stressor

It is well known that organisms do adapt to chronic stressors. This can take the form of phenotypic physiological adaptations that develop fairly quickly and alterations in the gene pool that require multiple generations. For example, in Virginia there is a river in which part has been contaminated with mercury for over 40 years. We were unable to detect important physiological differences in fish above and within the contaminated area even though the fish from the contaminated area had a considerable body burden of mercury and those above did not (Bidwell and Heath 1993). We are not claiming there is no problem here, because people do eat these fish, but apparently the mercury is having little detectable impact on the resident fish.

Determining Whether or Not We Are Exceeding the Assimilative Capacity of an Environment

Ecosystem assimilation capacity assumes that chemicals can be added to a body of water up to some point above which harm will be done. This concept is controversial (Cairns 1989) in part because response thresholds are always difficult to define. Odum (1984) has hypothesized that natural systems under stress may become more variable, which would make the determination of no-effect thresholds even more difficult to determine. However, if there are certain species of interest from an economic or esthetic standpoint, physiological measurements on these residents can help reveal whether the anthropogenic effects are causing measurable harm to them.

REFERENCES

Adams, S. M. and McLean, R. B., Estimation of largemouth bass, *Micropterus salmoides,* growth using the liver somatic index and physiological variables, *J. Fish Biol.,* 26, 111, 1985.

Adams, S. M. and Breck, J. E., Bioenergetics, in *Methods for Fish Biology,* Schreck, C. B. and Moyle, P. B., Eds., American Fisheries Society, Bethesda, MD, 1990, chap. 12.

Adams, S. M., Crumby, W. D., Greeley, M., Ryon, M. G., and Schilling, E. M., Relationship between physiological and fish population responses in a contaminated stream, *Environ. Toxicol. Chem.,* 11, 1549, 1992.

Anderson, D. P., Immunological indicators: Effects of environmental stress on immune protection and disease outbreaks, *American Fisheries Society Symposium,* 8, 38, 1990.

Anderson, D. E., Reid, S. C., Moon, T. W., and Perry, S. F., Metabolic effects associated with chronically elevated cortisol in rainbow trout *(Oncorhynchus mykiss), Can. J. Fish. Aquat. Sci.,* 48, 1811, 1991.

Barron, M. G. and Adelman, I. R., Nucleic acid, protein content, and growth of larval fish sublethally exposed to various toxicants, *Can. J. Fish. Aquat. Sci.,* 41, 141, 1984.

Barry, T. P., Lapp, A. F., Kayes, T. B., and Malison, J. A., Validation of a microtitre plate ELISA for measuring cortisol in fish and comparison of stress responses of rainbow trout, *Oncorhynchus mykiss,* and lake trout, *Salvelinus namaycush, Aquaculture,* 117, 351, 1993.

Barton, B. A. and Iwama, G. K., Physiological changes in fish from stress in aquaculture with emphasis on the response and effects of corticosteroids, *Ann. Rev. Fish Dis.,* 1, 3, 1991.

Bass, M. L. and Heath, A. G., Cardiovascular and respiratory changes in rainbow trout, *Salmo gairdneri,* exposed intermittently to chlorine, *Water Res.,* 11, 497, 1977.

Beamish, F. W. H., Swimming capacity, in *Fish Physiology,* Volume 7, Hoar, W. S. and Randall, D. J., Eds., Academic Press, New York, 1978, chap. 2.

Beitinger, T., Behavioral reactions for the assessment of stress in fishes, *J. Great Lakes Res.,* 16, 495, 1990.

Bidwell, J. R. and Heath, A. G., An *in situ* study of rock bass (*Ambloplites rupestris*) physiology: effect of season and mercury contamination, *Hydrobiologia,* 264, 137, 1993.

Binder, R. L. and Lech, J. J., Xenobiotics in gametes of Lake Michigan lake trout *Salvelinus namaycush* induce hepatic monooxygenase activity in their offspring, *Fund. Appl. Toxicol.,* 4, 1042, 1984.

Black, J. A. and Birge, W. J., An avoidance response bioassay for aquatic pollutants, Research Report No. 123, Univ. Kentucky, Water Resource. Res. Inst., Lexington, Ky., 1980.

Blaxhall, P. C. and Daisley, K. W., Routine haematological methods for use with fish blood, *J. Fish Biol.,* 5, 771, 1973.

Blaxter, J. and Hallers-Tjabbes, C. T., The effect of pollutants on sensory systems and behaviour of aquatic animals, *Netherlands J. Aquat. Ecol.,* 26, 43, 1992.

Bouck, G. R., Physiological responses in fish: Problems in progress toward use in environmental monitoring, in *Contaminant Effects on Fisheries,* Cairns, V. W., Hodson, P. V., and Nriagu, J. O., Eds., John Wiley & Sons, New York, 1984, chap. 6.

Bradley, B. P., Are the stress proteins indicators of exposure or effect?, *Mar. Environ. Res.*, 35, 85, 1993.

Brett, J. R. and Groves, T. D., Physiological energetics, in *Fish Physiology*, Vol. VIII, Hoar, W. and Randall, D., Eds., Academic Press, New York, 1979, chap. 6.

Buckley, L. J., Halavik, T., Laurence, G. C., Hamilton, S. J., and Yevitch, P., Comparative swimming stamina, biochemical composition, backbone mechanical properties and histopathology of juvenile striped bass from rivers and hatcheries of the Eastern United States, *Trans. Am. Fish. Soc.*, 114, 114, 1985.

Bulow, F. J., RNA:DNA ratios as indicators of growth in fish: a review, in *Age and Growth of Fish*, Summerfelt, R. C. and Hall, G. E., Eds., Iowa State University Press, Ames, 1987, 45.

Busacker, G. P., Adelman, I. R., and Goolish, E. M., Growth, in *Methods for Fish Biology*, Schreck, C. B. and Moyle, P. B., Eds., American Fisheries Society, Bethesda, 1990, 363.

Cairns, J., Applied ecotoxicology and methodology, in *Aquatic Ecotoxicology: Fundamental Concepts and Methodologies*, Boudou, A. and Ribeyre, F., Eds., CRC Press, Boca Raton, 1989, chap. 6.

Casillas, E., Myers, M., Rhodes, L., and McCain, B. B., Serum chemistry of diseased English sole, *Parophrys vetulus*, from polluted areas of Puget Sound, Washington, *J. Fish Dis.*, 8, 437, 1985.

Coppage, D. O. and Mathews, E., Short-term effects of organophosphate pesticides on cholinesterases of estuarine fishes and pink shrimp, *Bull. Environ. Contam. Toxicol.*, 11, 483, 1974.

Davis, K. B., Torrance, P., Parker, N., and Shuttle, M., Growth, body composition and hepatic tyrosine aminotransferase activity in cortisol-fed channel catfish, *Ictalurus punctatus*, *J. Fish Biol.*, 27, 177, 1985.

Dixon, D. G., Hill, C. E. A., Hodson, P. V., Kempe, E. J., and Kaiser, K. L. E., Plasma leucine aminonaphthylamidase as an indicator of acute sublethal toxicant stress in rainbow trout, *Environ. Toxicol. Chem.*, 4, 789, 1985.

Dixon, D. G., Hodson, P. V., and Kaiser, K. L., Serum sorbitol dehydrogenase activity as an indicator of chemically induced liver damage in rainbow trout, *Environ. Toxicol. Chem.*, 6, 685, 1987.

Donaldson, E. M., The pituitary-interrenal axis as an indicator of stress in fish, in *Stress and Fish*, Pickering, A. D., Ed, Academic Press, New York, 1981, chap. 2.

Engel, D. W. and Brouwer, M., Metallothionein and metallothionein-like proteins: Physiological importance, *Adv. Comp. Environ. Physiol.*, 4, 53, 1989.

Evans, D., Ed., *The Physiology of Fishes*, CRC Press, Boca Raton, 1993a.

Evans, D., Osmotic and ionic regulation, in *The Physiology of Fishes*, CRC Press, Boca Raton, 1993b, chap. 11.

Evans, D., Ed., *The Physiology of Fishes*, 2nd Ed., CRC Press, Boca Raton, FL, 1997.

Everhaarts, J., Shugart, L., Gustin, M., Hawkins, W., and Walker, W., Biological markers in fish: DNA integrity, hematological parameters and liver somatic index, *Mar. Environ. Res.*, 35, 101, 1993.

Faisal, M., Weeks, B. A., Vogelbein, W. K., and Huggett, R. J., Evidence of aberration of the natural cytotoxic cell activity in *Fundulus heteroclitus* (Pisces: Cyprinodontidae) from the Elizabeth River Virginia, *Veterinary Immunol. Immunopath.*, 29, 339, 1991.

Fausch, K. D., Lyons, J., Karr, J. R., and Angermeier, P., Fish communities as indicators of environmental degradation, *Am. Fish. Soc. Symp.*, 8, 123, 1990.

Folmar, L. C., Effects of chemical contaminants on blood chemistry of teleost fish: a bibliography and synopsis of selected effects, *Environ. Toxicol. Chem.*, 12, 337, 1993.

Forlin, L. and Celander, M., Induction of cytochrome P450-A in teleosts: environmental monitoring in Swedish fresh, brackish and marine waters, *Aquatic Toxicol.*, 26, 41, 1993.

Freadman, M. A., Thurberg, F. P., and Calabrese, A., Swimming/locomotor capacity of Hudson River striped bass, in *Marine Pollution and Physiology: Recent Advances,* Vernberg, F. J., Ed., Univ. South Carolina Press, Columbia, 1985, 31.

Gantverg, A. N. and Perevoznikov, M. A., Inhibition of cholinesterase in the brain of perch, *Perca fluviatilis* (Percidea) and common carp, *Cyprinus carpio* (Cyprinidae) under action of carbophos, *J. Ichthyology,* 23, 174, 1984.

Giesy, J. P., Phosphoadenylate concentrations and adenylate energy charge of largemouth bass (*Micropterus salmoides*): relationship with condition factor and blood cortisol, *Comp. Biochem. Physiol.*, 90A, 367, 1988.

Goede, R. W. and Barton, B. A., Organismic indices and an autopsy-based assessment as indicators of health and condition of fish, *Am. Fish. Soc. Symp.*, 8, 93, 1990.

Goksoyr, A. and Forlin, L., The cytochrome P-450 system in fish, aquatic toxicology and environmental monitoring, *Aquatic Toxicol.*, 22, 287, 1992.

Gray, R. H., Fish behavior and environmental assessment, *Environ. Toxicol. Chem.*, 9, 53, 1990.

Grizzle, J. M., Horowitz, S. A., and Strength, D. R., Caged fish as monitors of pollution: effects of chlorinated effluent from a wastewater treatment plant, *Water Resources Bull.*, 24, 951, 1988.

Gruber, D. J., Diamond, J. M., and Parsons, M. J., Automated biomonitoring, *Environmental Auditor,* 2, 229, 1991.

Guth, G. and Hanke, W., A comparison of physiological changes in carp, *Cyporinus carpio,* induced by several pollutants at sublethal concentrations I. The dependency on exposure time, *Ecotox. Environ. Safety,* 9, 179, 1985.

Haasch, M. L., Prince, R., Wejksnora, P. J., Cooper, K. R., and Lech, J. J., Caged and wild fish: induction of hepatic cytochrome P-450 (CYP1A1) as an environmental monitor, *Env. Toxicol. Chem.*, 12, 885, 1993.

Halver, J. E., Smith, R. R., Tolbert, B. M., and Baker, E. M., Utilization of ascorbic acid in fish, *Ann. N. Y. Acad. Sci.*, 258, 81, 1975.

Hamer, D. H., Metallothioneins, *Ann Rev. Biochem.*, 55, 913, 1986.

Hamilton, S. J. and Mehrle, P. M., Metallothionein in fish: review of its importance in assessing stress from metal contaminants, *Trans. Am. Fish. Soc.*, 115, 596, 1986.

Haux, C., Larsson, A., Lithner, G., and Sjobeck, M., A field study of physiological effects on fish in lead-contaminated lakes, *Env. Toxicol. Chem.*, 5, 283, 1985a.

Haux, C., Maj-Lis Sjobeck, and Larsson, A., Physiological stress responses in a wild fish population of perch (*Perca fluviatilis*) after capture and during subsequent recovery, *Mar. Environ. Res.*, 15, 77, 1985b.

Heath, A. G., Changes in tissue adenylates and water content of bluegill, *Lepomis macrochirus,* exposed to copper, *J. Fish Biol.*, 24, 299, 1984.

Heath, A. G., *Water Pollution and Fish Physiology,* 2nd Ed., CRC Press, Boca Raton, 1995.

Heath, A. G., Summary and perspectives, *American Fisheries Society Symposium*, 8, 183, 1990.

Heath, A. G., Cech, J. J., Zinkl, J. G., Finlayson, B., and Fujimura, R., Sublethal effects of methyl parathion, carbofuran, and molinate on larval striped bass, *Am. Fish. Soc. Symp.*, 14, 17, 1993.

Hille, S., A literature review of the blood chemistry of rainbow trout, *Salmo gairdneri* Rich., *J. Fish Biol.*, 20, 535, 1982.

Holeton, G. F., Gas exchange in fish with and without hemoglobin, *Respir. Physiol.*, 14, 142, 1972.

Hontela, A., Rasmussen, J. B., Audet, C., and Ghevalier, G., Impaired cortisol stress response in fish from environments polluted by PAHs, PCBs and mercury, *Arch. Environ. Contam. Toxicol.*, 22, 278, 1992.

Houlihan, D. F., Mathers, E. M., and Foster, A., Biochemical correlates of growth rate in fish, in *Fish Ecophysiology*, Rankin, J. C. and Jensen, F. B., Eds., Chapman & Hall, London, 1993, chap. 2.

Houston, A., Blahut, S., Murad, A., and Amirtharaj, P., Changes in erythron organization during prolonged cadmium exposure: An indicator of heavy metal stress?, *Can. J. Fish. Aquat. Sci.*, 50, 217, 1993.

Jobling, M., Bioenergetics: feed intake and energy partitioning, in *Fish Ecophysiology*, Rankin, J. C. and Jensen, F. B., Eds., Chapman & Hall, London, 1993, chap. 1.

Kohler, A., Lysosomal perturbations in fish liver as indicators for toxic effects of environmental pollution, *Comp. Biochem. Physiol.*, 100C, 123, 1991.

Larsson, A., Andersson, T., Forlin, L., and Hardig, J., Physiological disturbances in fish exposed to bleached kraft mill effluents, *Wat. Sci. Tech.*, 20, 67, 1988.

Lee, R., Gerking, S., and Jezierska, B., Electrolyte balance and energy mobilization in acid-stressed rainbow trout, *Salmo gairdneri*, and their relation to reproductive success, *Environ. Biol. Fish.*, 8, 115, 1983.

Lindstrom-Seppa, P. and Oikari, A., Biotransformation and other toxicological and physiological responses in rainbow trout (*Salmo gairdneri*) caged in a lake receiving effluents of pulp and paper industry, *Aquat. Toxicol.*, 16, 187, 1990.

Little, E. E., Fairchild, J., and DeLonay, A., Behavioral methods for assessing impacts of contaminants on early life stage fishes, *Am. Fish. Soc. Symp.*, 14, 67, 1993.

Little, E. E., Archeski, R. D., Flerov, B. A., and Kozlovskaya, V. I., Behavioral indicators of sublethal toxicity in rainbow trout, *Arch. Environ. Contam. Toxicol.*, 19, 380, 1990.

Livingston, D. R., Biochemical measures, in *The Effects of Stress and Pollution on Marine Animals,* Bayne, B. L., Brown, D. A., Burns, K., Dixon, D. R., Ivanovici, A., Livingstone, D., Lowe, D., Moore, M., Stebbing, A. R. D., and Widdows, J., Eds., Praeger Publishers, New York, 1985, 81.

Mallatt, J., Fish gill structural changes induced by toxicants and other irritants; a statistical review, *Can. J. Fish. Aq. Sci.*, 42, 630, 1985.

Mathers, E. M., Houlihan, D. F., and Cunningham, M. J., Nucleic acid concentrations and enzyme activities as correlates of growth rate of the saithe, *Pollachius virens:* growth-rate estimates of open-sea fish, *Marine Biology,* 112, 363, 1992.

Mayer, F. L., Versteeg, D. J., McKee, M. J., Folmar, L. C., Graney, R. L., McCume, D. C., and Rattner, B. A., Physiological and nonspecific biomarkers, in *Biomarkers: Biochemical, Physiological and Histological Markers of Anthropogenic Stress*, Huggett, R. J., Kimerle, R. A., Mehrle, P. M., and Bergman, H. L., Eds., Lewis Publishers, Boca Raton, 1992, chap. 1.

McKim, J. M. and Goeden, H. M., A direct measure of the uptake efficiency of a xenobiotic chemical across the gills of brook trout (*Salvelinus fontinalis*) under normoxic and hypoxic conditions, *Comp. Biochem. Physiol.*, 72C, 65, 1982.

Melancon, M. J., Yeo, S. E., and Lech, J. J., Induction of hepatic microsomal monooxygenase activity in fish by exposure to river water, *Environ. Toxicol. Chem.*, 6, 127, 1987.

Melancon, M. J., Binder, R. L., and Lech. J. J., Environmental induction of monooxygenase activity in fish, in *Toxic Contaminants and Ecosystem Health,* A Great Lakes Focus, Evans, M. S., Ed., John Wiley & Sons, New York, 1988, 215.

Miller, W. R., Hendricks, A. C., and Cairns, J., Normal ranges for diagnostically important hematological and blood chemistry characteristics of rainbow trout (*Salmo gairdneri*), *Can. J. Fish. Aquat. Sci.*, 40, 420, 1983.

Mineau, P., Ed., *Cholinesterase-inhibiting Insecticides, Their Impact on Wildlife and the Environment,* Elsevier, Amsterdam, 1991.

Morgan, E. L., Young, R. C., and Wright, J. R., Developing portable computer-automated biomonitoring for a regional water quality surveillance network, in *Automated Biomonitoring,* Gruber, D. J. and Diamond, J. M., Eds., Halsted Press, London, 1988, chap. 9.

NRDA, Natural Resource Damage Assessments: Final Rule, *Federal Register,* 51, 27674, 1986.

Niimi, A. J. and Lowe-Jinde, L., Differential blood cell ratios of rainbow trout (*Salmo gairdneri*) exposed to methylmercury and chlorobenzenes, *Arch. Environ. Contam. Toxicol.,* 13, 303, 1984.

O'Connor, J. S., Ziskowski, J. J., and Murchelano, R. A., Index of pollutant-induced fish and shellfish disease, NOAA (National Oceanic and Atmospheric Administration) Special Report, Washington, D.C., 1987.

Odum, E. P., The mesocosm, *Bioscience,* 34, 558, 1984.

Payne, J. F., Fancey, L. L., Rahimtula, A. D., and Porter, E. L., Review and perspective on the use of mixed-function oxygenase enzymes in biological monitoring, *Comp. Biochem, Physiol.,* 86C, 233, 1987.

Peterson, R. H., Influence of fenitrothion on swimming velocity of brook trout *(Salvelinus fontinalis), J. Fish Res. Bd. Can.,* 31, 1757, 1974.

Pickering, A. D., Pottinger, T. G., and Christie, P., Recovery of the brown trout, *Salmo trutta,* from acute handling stress: a time-course study, *J. Fish Biol.,* 20, 229, 1982.

Pickering, A. D. and Pottinger, T., Stress responses and disease resistance in salmonid fish: effect of chronic elevation of plasma cortisol, *Fish Physiol. Biochem.,* 7, 253, 1989.

Ram, R. and Singh, R. S., Carbofuran-induced histopathological and biochemical changes in liver of the teleost fish, *Channa punctatus, Ecotox. Environ. Safety,* 16, 194, 1988.

Rankin, J. C. and Jensen, F. B., Eds., *Fish Ecophysiology,* Chapman & Hall, London, 1993.

Rao, D. P., Bhaskar, B. R., Rao, K. S., Prasad, Y. V. K., Rao, N. S., and Rao, T. N. V. V., Hematological effects in fishes from complex polluted waters of Visakhapatnam harbour, *Mar. Environ. Res.,* 30, 217, 1990.

Roesijadi, G., Metallothioneins in metal regulation and toxicity in aquatic animals, *Aquat. Toxicol.,* 22, 81, 1992.

Sanders, B. M., Stress proteins in aquatic organisms: an environmental perspective, *Crit. Rev. Toxicol.,* 23, 49, 1993.

Schreck, C. B. and Lorz, H. W., Stress response of coho salmon (*Oncorhynchus kisutch*) elicited by cadmium and copper and potential use of cortisol as an indicator of stress, *J. Fish. Res. Bd. Can.,* 35, 1124, 1978.

Shaw, B. P. and Panigraphi, A. K., Brain AChE activity studies in some fish species collected from a mercury contaminated estuary, *Water Air Soil Poll.,* 53, 327, 1990.

Sindermann, C. J., Pollution-associated diseases and abnormalities of fish and shellfish: a review, *Fish. Bull.,* 76, 717, 1979.

Sloof, W., van Kreijal, C., and Baars, A., Relative liver weights and xenobiotic-metabolizing enzymes of fish from polluted surface waters in the Netherlands, *Aquat. Toxicol.,* 4, 1, 1983.

Sprague, J. B., Measurement of pollutant toxicity to fish — III. Sublethal effects and "safe" concentrations, *Water Research,* 5, 245, 1971.

Stegeman, J. J., Brouwer, M., Di Giulio, R. T., Forlin, L., Fowler, B., Sanders, B., and Van Veld, P. A., Molecular responses to environmental contamination: enzyme, and protein systems as indicators of chemical exposure, in *Biomarkers: Biochemical, Physiological, and Histological Markers of Anthropogenic Stress,* Huggett, R. J., Kimerle, R. A., Mehrle. P. M., and Bergman, H. L., Eds., Lewis Publishers, Boca Raton, 1992, chap. 6.

Stegeman, J. J. and Hahn, M. E., Biochemistry and molecular biology of monooxygenases: current perspectives on forms, functions and regulation of cytochrome P450 in aquatic species, in *Aquatic Toxicology, Molecular, Biochemical and Cellular Perspectives,* Malins, D. C. and Ostrander, G. K., Eds., CRC Press, Boca Raton, 1994, chap. 3.

Stein, J. E., Collier, T. K., Reichert, W. L., Casillas, E., Hom, T., and Varanasi, U., Bioindicators of contaminant exposure and sublethal effects: studies with benthic fish in Puget Sound, Washington, *Environ. Toxicol. Chem.,* 11, 701, 1992.

Suter, G. W., A critique of ecosystem health concepts and indexes, *Environ. Toxicol. Chem.,* 12, 1533, 1993.

Theodorakis, C. W., Surney, S. J., Bickham, J. W., Lyne, T. B., Bradley, B. P., Hawkins, W. E., Farkas, W. L., McCarthy, J. F., and Shugart, L. R., Sequential expression of biomarkers in bluegill sunfish exposed to contaminated sediment, *Ecotoxicology,* 1, 45, 1992.

Thomas, P., Molecular and biochemical responses of fish to stressors and their potential use in environmental monitoring, *Am. Fish. Soc. Symp.,* 8, 9, 1990.

Thomas, P. and Juedes, M. J., Influence of lead on the glutathione status of Atlantic croaker tissues, *Aquat. Toxicol.,* 23, 1992.

Thomas, P. and Lewis, D. H., Effects of cortisol on immunity in red drum, *Sciaenops ocellatus, J. Fish Biol.,* 31 (Supp. A.), 123, 1987.

Thomas, P., Bally, M., and Neff, J., Ascorbic acid status of mullet, *Mugil cephalus,* exposed to cadmium, *J. Fish Biol.,* 20, 183, 1982.

Thomas, P. and Neff, J., Effects of pollutant and other environmental variables on the ascorbic acid content of fish tissues, *Mar. Environ. Res.,* 14, 489, 1984.

Thomas, P. and Wofford, H. W., Effects of metals and organic compounds on hepatic glutathione, cysteine, and acid-soluble thiol levels in mullet, *Mugil cephalus, Toxicol. Appl. Pharmacol.,* 76, 172, 1984.

Tytler, P. and Calow, P., Eds., *Fish Energetics, New Perspectives,* Johns Hopkins Press, Baltimore, 1985.

Ueberschar, D. F. R., Determination of the nutritional condition of individual marine fish larvae by analyzing their proteolytic enzyme activities with a highly sensitive fluorescence technique, *Meeresforschung,* 32, 144, 1988.

Van der Boon, J., van den Thillart, G., and Addink, A. D. F., The effects of cortisol administration on intermediary metabolism in teleost fish, *Comp. Biochem. Physiol.,* 100B, 47, 1991.

Verma, S. R., Tonk, I. P., and Dalela, R. C., Effects of a few xenobiotics on three phosphatases of *Saccobranchus fossilis* and the role of ascorbic acid in their toxicity, *Toxicity Letters,* 10, 287, 1982.

Versteeg, D. and Giesy, J. P., Lysosomal enzyme release in the bluegill sunfish (*Lepomis macrochirus*) exposed to cadmium, *Arch. Environ. Contam. Toxicol.,* 14, 631, 1985.

Vignier, V., Vandermeulen, J., and Fraser, A., Growth and food conversion by Atlantic salmon parr during 40 days exposure to crude oil, *Trans. Am. Fish. Soc.,* 121, 322, 1992.

Waldichuk, M., Review of the problems, in *Assessment of Sublethal Effects of Pollutants in the Sea,* Cole, H. A., Ed., *Phil. Trans. R. Soc. London,* B, 286, 397, 1979.

Weeks, B. A., Anderson, D. P., DuFour, A. P., Fairbrother, A., Goven, A. J., Lahvis, G. P., and Peters, G., Immunological biomarkers to assess environmental stress, in *Biomarkers, Biochemical, Physiological, and Histological Markers of Anthropogenic Stress,* Huggett, R. J., Kimerle, R. A., Mehrle, P. M., and Bergman, H. L., Eds., Lewis Publishers, Boca Raton, 1992, chap. 5.

Welch, W. J., How cells respond to stress, *Scientific Am.,* May, 56, 1993.

Wendelaar Bonga, S. E. and Lock, R. A. C., Toxicants and osmoregulation in fish, *Netherlands J. Zool.,* 42, 478, 1992.

Wilson, R. W. and Wood, C. M., Swimming performance, whole body ions, and gill Al accumulation during acclimation to sublethal aluminum in juvenile rainbow trout (*Oncorhynchus mykiss), Fish Physiol. Biochem.,* 10, 149, 1992.

Wood, C. M., The physiological problems of fish in acid waters, in *Acid Toxicity and Aquatic Animals,* Morris, R., Taylor, E., Brown, D., and Brown, J., Eds., Cambridge University Press, New York, 1989, 125.

Yamamato, Y., Hayama, K., and Ikeda, S., Effect of dietary ascorbic acid on the copper poisoning in rainbow trout, *Bull. Jpn. Soc. Sci. Fish.*, 47, 1085, 1981.

Zinkl, J., Lockhart, W., Kenny, S., and Ward, F., Effects of cholinesterase-inhibiting insecticides on fish, in *Cholinesterase-Inhibiting Insecticides — Impact on Wildlife and the Environment*, Mineau, P., Ed., Elsevier Science Publishers, Amsterdam, 1991, 233.

8 Biomarkers: Are There Linkages to Ecological Effects?

Pierre Mineau

INTRODUCTION

Because the title of this section ("Establishing the Health of Ecosystems") is rather forbidding, perhaps I should start with what I do *not* intend to cover in my chapter. I will not engage in the continuing debate of whether human health and homeostasis are useful metaphors of ecosystem health (Suter 1993 vs. Rapport 1989). I will also not discuss whether the health of an ecosystem can be reduced to a single index (Karr 1993), needs to be described as a complex multivariate space (Suter 1993), or can even be defined given the usual imprecisions of the meaning of both "health" and "ecosystem."

Even though I will use the term, I prefer not to formally define what constitutes a healthy ecosystem. As mentioned by a number of participants in the symposium that inspired this book, concepts of ecosystem health, quality, or integrity are, at least in part, subjective. A common strategy is to define an unhealthy ecosystem as one which, by some objective measure, has deviated from its trajectory, has lost its ability to self-regulate, or no longer meets its full potential in terms of its biotic diversity or its productivity.

I shall use the narrow definition of biomarker: that is, a *physiological*, *biochemical*, or *histological* manifestation indicative of stress, toxic or otherwise. A biomarker as used here is therefore first and foremost a measure of individual health. The concept of biomarker has been extended by some to refer to functional aspects of ecosystems such as measures of diversity, connectivity, or efficiency. The usefulness of such ecological measurements in assessing ecosystem health is intuitively obvious. Could biomarkers of individual health be used to assess ecosystems? In other words, do healthy individuals signify that the ecosystem is healthy, and is the converse true?

I do not propose to offer a systematic or comprehensive review of available biomarkers: biomarkers of individual health have been extensively reviewed in a number of recent texts, some covering a single biomarker in depth (e.g., Mineau 1991a); others reviewing the field at large (e.g., McCarthy and Shugart 1990; Huggett et al. 1992; Peakall 1992; Peakall and Shugart 1993).

As witnessed by an ever-increasing number of international initiatives (the Brundtland Commission, the Rio Convention on Biodiversity, the Montreal Protocol

1-56670-309-3/98/$0.00+$.50

on chlorofluorocarbons etc.), there are increasing concerns over the deterioration of our planet. There also seems to be a consensus that we need better measuring tools to guide us in making serious (and often costly) decisions about the present and future quality of our environment. The question I have been asked to address is whether physiological, biochemical, or histological biomarkers have a role to play in this decision-making process.

SOME IMPORTANT CHARACTERISTICS OF BIOMARKERS

Biomarkers are acknowledged to be useful as (1) integrated measures of exposure, (2) replacements for chemical residue determinations, and (3) indicators of stress operating on the individual and, when compensatory mechanisms are overwhelmed, of impairment resulting from that stress. A good biomarker is generally recognized to have the following attributes:

1. Good signal-to-noise ratio in terms of its sensitivity to perturbation.
2. A short response time. Kelly and Harwell (1989) have argued that this is likely to occur at lower trophic levels, although there are well-documented examples (e.g., the bioaccumulation of persistent organohalides) where higher trophic levels provided the earliest indication of physiological or biochemical impairment.
3. Specificity of response. This point has probably been the most debated. In a framework of single stressor management, a high degree of response specificity in a biomarker is critical in order to make appropriate management decisions. For example, injury from known contaminants should not be mistaken for injury resulting from environmental perturbations such as climatic or food stress if the goal is to decide whether environmental loadings of the contaminants in question need to be controlled. On the other hand, it can be and has been argued that nonspecific biomarkers (e.g., immune function, condition indices etc.) offer a better assessment of the effect of multiple stressors (both "natural" and anthropogenic) acting in concert on the individual. A common strategy here is to look at a battery of such biomarkers across a gradient of sites ecologically comparable but differing in the anthropogenic factor of interest, e.g., exposure to contaminants (see Fox 1993). Unfortunately, using nonspecific biomarkers means that cause–effect relationships are often inferential and open to debate.
4. A quantifiable response which provides a measure of the degree of stress, impairment, or injury.
5. Ease and economy of measurement.
6. Given the importance currently placed on conservation of biological diversity (UNEP 1992), we could also add another desirable feature: that biomarkers provide us with a measure of the loss of genetic information from any given species.

In the field of environmental contamination, biomarkers are often used for the assessment of human health or as a measure of human exposure (either directly or through surrogate species). For our expressed purpose of measuring the health of ecosystems, however, the biomarker should provide information about other important parts of the ecosystem under consideration. A species may be deemed important for our narrow utilitarian or aesthetic reasons, but more important, it may play a key role in the ecosystem. Such species are usually referred to as "keystone" species, those on which much of the current ecosystem structure depends. Clearly, if the goal of measuring individual health and impairment is to make inferences about the ecosystem at large, then the indicator species need to be chosen carefully and reflect a broad range of ecosystem levels and functions.

CURRENT VIEWS ON THE RELATIONSHIP BETWEEN BIOMARKERS AND ECOSYSTEM HEALTH

One of the conclusions of the NATO Expert Meeting on biomarkers (Peakall and Shugart 1993) was that, given our current state of knowledge, biomarkers, as defined above, are unlikely to shed light on ecosystem-level properties. Our ability to predict from the cellular to the individual is far from perfect and, realistically, predicting from the cellular or physiological to the population or ecosystem may be an unattainable goal. This is a restatement of what has been referred to as Fry's paradigm: in order to understand a system, one must study properties of that system rather than properties of its subcomponents (Kerr 1976).

Notwithstanding the possibility that the species under observation is a keystone species, it is difficult to argue that even major impacts on a few species will jeopardize the functioning of an ecosystem. This is because of the generally recognized degree of redundancy in most ecosystems. As pointed out by some (e.g., Ehrenfeld 1988; King 1993), there are examples of ecosystems which have seemingly withstood the extirpation of once common or dominant species. The disappearance of the American chestnut (*Castanea dentata*) and passenger pigeon (*Ectopistes migratorius*) from eastern North American forest systems are such examples. The ecosystem did not grind to a halt but merely shifted and trundled along.

The question is therefore whether we can still use physiological or biochemical biomarkers measured on individuals to help us manage ecosystems, even if those biomarkers do not clearly give us an objective measure of ecosystem health. I will focus the rest of this chapter on one response-specific biomarker and provide an example where this biomarker was used to achieve regulatory relief in a context of prudent environmental management in Canada.

MEASURING CHOLINESTERASE ACTIVITY: A CASE STUDY OF AN IDEAL BIOMARKER

Biomarkers can be ranked for their specificity and level of injury indicated. Measuring the inhibition of cholinesterase (ChE) activity is a biomarker of very high specificity. This biomarker is highly indicative of exposure to cholinesterase-inhibiting agents such as organophosphorous (OP) or carbamate pesticides. The measurement of ChE inhibition is inexpensive and highly reproducible. The same method works over a very wide range of biota, and the use of this biomarker in assessing pesticide hazard has been extensively reviewed and validated (e.g., Mineau 1991a). It is one of the few biomarkers formally accepted to indicate environmental damage under U.S. CERCLA legislation (Comprehensive Environmental Response, Compensation, and Liability Act).

Acetylcholinesterase (ACHE) is needed in nervous tissue and neuromuscular junctions to catalyze the neurotransmitter acetylcholine. Acetylcholine is ubiquitous; it is the only known substance that can influence practically every physiological or behavioral response examined. More important, the inhibition of ACHE is a meaningful response in that, if excessive, it leads to death. Although the exact relationship between expression of the biomarker and the actual impact on the organism can presently only be described in probabilistic terms, especially at lower levels of inhibition, it is probably better defined than for most other biomarkers in wide use today. The inhibition of ACHE (often measured as total ChE) is therefore an ideal biomarker that every ecotoxicologist dreams of having at his or her disposal.

Because the ChE biomarker is nearly specific to OP and carbamate compounds, ecosystems where these chemicals are heavily used are prime targets. Two such situations are given below.

The Use of ChE Inhibition to Assess the Health of Agricultural Landscapes

Agricultural ecosystems (defined here as landscapes dominated by agricultural activity) are prime examples of systems under multiple stress. Some of these stresses are:

- A lack of suitable habitat for natural biota resulting from cumulative habitat loss and severe fragmentation of any remaining habitat (Ratti and Scott 1991).
- A high disturbance regime (including tillage, harvest, grazing, soil compaction, erosion) which has a negative impact on the conservation of native species assemblages (McLaughlin and Mineau 1995).
- Exposure to surplus levels of N, P, as well as a number of trace metals (van Monsjou 1975).
- Exposure to massive numbers of exotic species (Mulligan 1965).
- Repeated exposure to a wide range of biocides — including the two pesticide classes for which this biomarker is useful (OPs and carbamates).

There is a fairly generalized opinion that, as a result of the factors enumerated above (and others), modern agricultural ecosystems are inherently unhealthy or, to use a currently popular label: unsustainable (World Commission on Environment and Development 1987; National Academy of Sciences 1989). They require large, repeated energy inputs; they export contaminants to other systems; they slowly lose their carbon reserve over time; they usually support a low level of native biotic diversity and are dominated by species exotic to the region, many of which are pests and need to be actively controlled.

Unfortunately, our biomarker tells us almost nothing about these all-important ecological considerations. Its only use would be to assess whether vertebrate species have been exposed to and are likely to be affected by OP and carbamate insecticides. [The relationship between cholinesterase inhibition and biological impairment in aquatic and terrestrial invertebrates is so variable that this biomarker has not been used extensively to monitor the health of invertebrates (Edwards and Fisher 1991).] In truth, wild vertebrates probably play a minimal role in the functioning of present-day agricultural systems, and if they did, would probably be regarded as pests. The important ecosystem components are the soil community, above-ground arthropods, and plant life. We would have to conclude that measurements of ACHE activity, and indeed other biomarkers of vertebrate health and well being, tell us very little of the health of agricultural systems. At best, our biomarker may tell us whether vertebrate species inhabiting agricultural systems are likely to be impacted but not whether agricultural systems can even support vertebrate species, assuming this is a desirable goal. Cholinesterase-inhibiting insecticides might be replaced by other classes of insecticides (e.g., synthetic pyrethroids) or by increased tillage but, depending on the system and on the species, there may not be a measurable improvement in the capacity of the system to support vertebrates.

The Use of Cholinesterase Inhibition in Assessing the Health of Forest Ecosystems

Similarly in eastern Canada and the U.S., we have historically mounted extensive pest control programs to reduce the amount of natural timber loss to the primary insect defoliator, the Eastern Spruce Budworm (*Choristoneura fumiferana*). In Canada alone, annual spray programs can be in the millions of hectares (Busby et al. 1989). Over the last few decades, and until quite recently, the product of choice in eastern Canada was fenitrothion, a nonselective organophosphate insecticide. Given the lack of target specificity of this product, one would expect pronounced disruptions of the forest ecosystem in the spray areas, at least in the short term. Indeed, available evidence points to impacts on different segments of the forest environment such as aquatic and terrestrial nontarget invertebrates, songbirds, and possibly, amphibians (Pauli et al. 1993).

Yet, even though spraying on a large scale has been going on for decades, we would still be hard pressed today to say with any degree of certainty whether the forest system as a whole has been radically altered or is fundamentally unhealthy as a result of this practice. The necessary studies have not been carried out at a scale (geographical or temporal) commensurate with the problem. My guess is that in order to separate the effects of the spraying activity from (1) the vagaries of weather, (2) other sylvicultural manipulations, and (3) the impact of the pest itself on the complex forest ecosystem, we would need a massive concerted ecological effort (reminiscent of some IBP ecological programs of the 1970s but much more extensive) over a *minimum* of two or three budworm cycles (40 to 50 years). No one is

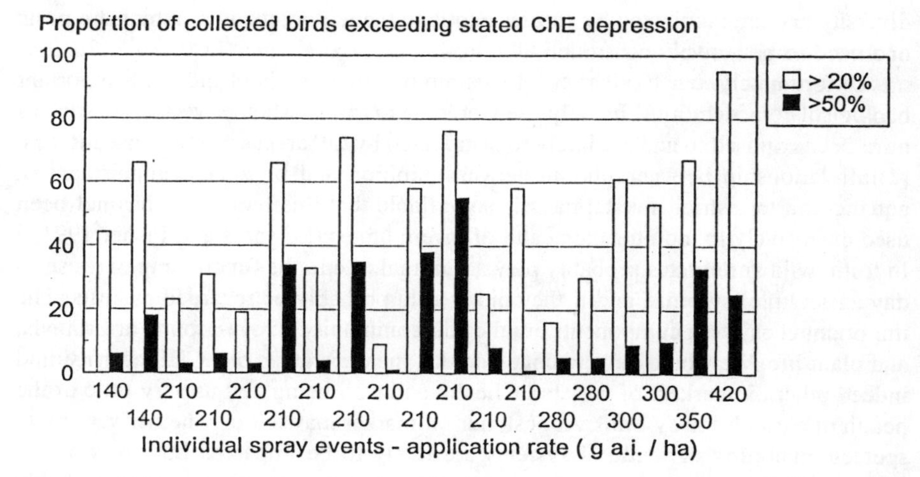

FIGURE 8.1 Proportion of collected forest songbirds exceeding either 20% or 50% inhibition in brain cholinesterase levels following exposure to fenitrothion. Each entry on the x axis represents a different spray event at the rate indicated in grams of active ingredient per hectare.

able to commit the required resources. There is even doubt in my mind as to whether we have the tools or the theoretical framework at our disposal to carry out such work.

Studies of the impact of fenitrothion in the forest ecosystem were restricted to a few groups. One of these was songbirds. This is one component of forest ecosystems that we value greatly. We have passed laws in North America protecting migratory songbirds at the level of the individual, and we carefully license any taking of these species. For the majority of songbirds (which were never directly threatened by human depredation), this is less a scientific position than a societal choice. Other groups (e.g., amphibians) were never afforded the same level of protection, although it might now be argued, in view of their declining populations, that this was a tragic oversight. There are now debates as to whether forest-inhabiting songbirds are in decline and, if they are, how much of this decline is due to human activities on their breeding grounds and how much is due to destruction of their neotropical winter range. We had no data to indicate whether fenitrothion was or was not affecting *population levels* of forest songbirds. Nevertheless, cholinesterase inhibition data collected in fenitrothion-sprayed areas (Figure 8.1) indicated that songbirds were massively exposed and probably frequently killed by applications of fenitrothion (Busby et al. 1989; Busby and White 1991; Pauli et al. 1993). Based on what we knew of the correlation between cholinesterase depression and possible consequences to the exposed individual, there was also a concern for exposed individuals even if they initially survived the exposure (Grue et al. 1991; Mineau 1991b).

On those grounds, the Canadian Wildlife Service of Environment Canada has had a long-standing opposition to this insecticide and, for a number of years, advised against the use of fenitrothion in favor of another insecticide, aminocarb, which did not result in a similar level of cholinesterase depression in songbirds. For a variety of reasons, aminocarb never became the product of choice.

There is now an alternative to fenitrothion which only affects a small proportion of the nontarget species affected by fenitrothion. The biological insecticide *Bacillus thuringiensis var. kurstaki* (B.t.) has been used by some jurisdictions to reduce spruce budworm defoliation to an economically acceptable level, although currently, its efficacy against the pest may be slightly lower and its use logistically more difficult. A full cost/benefit review of fenitrothion uses has just been concluded, and the product will be phased out and cease to be available for broad-scale spruce budworm control in Canada on December 31, 1998.

The opposition to, and replacement of, fenitrothion did not come as a result of data showing irremediable impacts on the health of eastern boreal forests. Instead, the position of the environmental managers was dictated by prudent pragmatism. The questions really boiled down to the following: Was it prudent to be using a broad-spectrum insecticide over large forested areas when a narrower-spectrum, less damaging product was available? Should we allow *needless* repeated impacts to protected songbirds?

CONCLUSION

In conclusion, biomarkers (defined as physiological, biochemical, or histological measures of health in individuals) are unlikely to tell us whether ecosystems are healthy or in danger of losing their integrity. They may tell us whether damage has occurred in a specific component of the ecosystem but not whether the ecosystem can compensate. Presumably, batteries of such biomarkers might be used to monitor several ecosystem components simultaneously, but realistically, questions relating to ecosystems will require that ecosystem-level measurements be made (Fry's paradigm). I suggest that biomarkers, especially those that are stressor-specific, will be used most effectively to trigger remediation measures before the health of ecosystems becomes an issue.

Ecosystems are shifting dramatically as we speak, whether as a result of introduced species, broad-scale contamination from anthropogenic substances, or simply from our direct influence on the landscape through destruction and loss of native habitat. Maintaining ecosystem health or integrity is like aiming at a shifting target. As argued by Levin (1989), ecosystem complexity in space and time is such that any characteristic of an ecosystem is not a property of that ecosystem as much as the interaction between that ecosystem and the observer. In other words, ecosystem complexity, and by extension ecosystem integrity or health, is largely a question of scale.

In my opinion, the most likely use for biomarkers is to measure conformity with a pre-defined standard of individual health. In the forestry example given here, the standard was, "Large numbers of forest songbirds should not be placed in jeopardy, because of depressed cholinesterase levels, especially when alternative methods of forest protection exist." The more ecologically defensible standard would have been, "Any kill of songbirds is acceptable provided population levels are not affected and the integrity of forest ecosystems is not jeopardized." No improvement of our forest pest control practices would have been possible under the second standard. There may not have been universal agreement that the first standard was the correct one, but it clearly was the prudent choice and the one most in keeping with our legal obligations toward these species. It is also clear that acceptable levels of species protection need not be fixed forever. Rather, they should improve in keeping with our increased technical sophistication. There will probably be a time when the use of B.t. for budworm control is unacceptable in view of more specific products which do not have the same impact on nontarget lepidopteran species.

Agricultural landscapes are characterized by a high degree of habitat manipulation, a relatively depauperate vertebrate fauna, and a high input of anthropogenic chemicals. My forestry colleagues may not appreciate the double standard, but I suspect it will be more difficult, in agricultural landscapes than in forest situations, to define unacceptable limits of exposure via measurements of cholinesterase activity in birds and other vertebrates. Up to now, even reports of repeated kills resulting from the use of cholinesterase-inhibiting insecticides in agricultural fields (Grue et al. 1983, and many others) have had remarkably little impact on the range of available products.

The challenge will be for our society to come to an agreement as to what constitutes an appropriate objective for any given ecosystem and whether, in the absence of detailed ecological information, biomarkers of individual health will be acceptable and, more important, actionable as *surrogates* for measures of ecosystem health.

ACKNOWLEDGMENTS

I am indebted to G.A. Fox, J.A. Keith, W.K. Marshall, and B. Pauli for comments on an earlier draft. I am also grateful to the conference organizers for the invitation to participate in this forum.

REFERENCES

Busby, D. G. and White, L. M., Factors influencing variability in brain acetylcholinesterase activity in songbirds exposed to aerial fenitrothion spraying, in *Cholinesterase Inhibiting Insecticides. Their Impact on Wildlife and the Environment*, Mineau, P., Ed., Elsevier, Amsterdam, 1991, chap. 9.

Busby, D. G., White, L. M., Pearce, P. A., and Mineau, P., Fenitrothion effects on forest songbirds: a critical new look, in *Environmental Effects of Fenitrothion Use in Forestry*, Ernst, W. B., Pearce, P. A., and Pollock, T. L., Eds., Environment Canada, Dartmouth N.S., 1989, chap. 3.

Edwards, C. A. and Fisher, S. W., The use of cholinesterase measurements in assessing the impacts of pesticides on terrestrial and aquatic invertebrates, in *Cholinesterase Inhibiting Insecticides. Their Impact on Wildlife and the Environment*, Mineau, P., Ed., Elsevier, Amsterdam, 1991, chap. 11.

Ehrenfeld, D., Why put a value on biodiversity?, in *Biodiversity*, Wilson, E. O., Ed., National Academy Press, Washington, D.C., 1988, chap. 24.

Fox, G. A., What have biomarkers told us about the effects of contaminants on the health of fish-eating birds in the Great Lakes? The theory and a literature review, *J. Great Lakes Res.*, 19, 722, 1993.

Grue, C. E., Fleming, W. J., Busby, D. G., and Hill, E. F., Assessing hazards of organophosphate pesticides to wildlife, *Trans. N. Am. Wildl. Nat. Resour. Conf.*, 48, 200, 1983.

Grue, C. E., Hart, A. D. M., and Mineau, P., Biological consequences of depressed brain cholinesterase activity in vertebrates, in *Cholinesterase Inhibiting Insecticides. Their Impact on Wildlife and the Environment*, Mineau, P., Ed., Elsevier, Amsterdam, 1991, chap. 8.

Huggett, R. J., Kimerie, R. A., Mehrle, P. M., Jr., and Bergman, H. L., *Biomarkers. Biochemical, Physiological, and Histological Markers of Anthropogenic Stress*, Lewis Publishers, CRC Press, Boca Raton, 1992.

Karr, J. R., Defining and assessing ecological integrity: Beyond water quality, *Environ. Toxicol. Chem.*, 12, 1521, 1993.

Kelly, J. R. and Harwell, M. A., Indicators of ecosystem response and recovery, in *Ecotoxicology: Problems and Approaches*, Levin, S. A., Harwell, M. A., Kelly, J. R., and Kimball, K. D., Eds., Springer-Verlag, New York, 1989, chap. 2.

Kerr, S. R., Ecological analysis and the Fry Paradigm, *J. Fish. Res. Bd. Can.*, 33, 329, 1976.

King, A. W., Considerations of scale and hierarchy, in *Ecological Integrity and the Management of Ecosystems*, Woodley, S., Kay, J., and Francis, G., Eds., St. Lucie Press, Boca Raton, 1993, 19.

Levin, S. A., Models in ecotoxicology: Methodological aspects, in *Ecotoxicology: Problems and Approaches*, Levin, S. A., Harwell, M. A., Kelly, J. R., and Kimball, K. D., Eds., Springer- Verlag, New York, 1989, chap. 8.

McCarthy, J. F. and Shugart, L. R., Eds., *Biomarkers of Environmental Contamination*, Lewis Publishers, CRC Press, Boca Raton, 1990.

McLaughlin, A. and Mineau, P., The impact of agricultural practices on biodiversity, *Agriculture, Ecosystems and Environment*, 55, 201, 1995.

Mineau, P., Ed., *Cholinesterase-inhibiting Insecticides. Their Impact on Wildlife and the Environment*, Elsevier, Amsterdam, 1991a.

Mineau, P., Difficulties in the regulatory assessment of cholinesterase-inhibiting insecticides, in *Cholinesterase Inhibiting Insecticides. Their Impact on Wildlife and the Environment*, Mineau, P., Ed., Elsevier, Amsterdam, 1991b, chap. 12.

Mulligan, G. A., Recent colonization by herbaceous plants in Canada, in *The Genetics of Colonizing Species*, Baker, H. G. and Stebbins, G. L., Eds., Academic Press, New York, 1965, 127.

National Academy of Sciences, *Alternative Agriculture*, National Academy Press, Washington, D.C., 1989.

Pauli, B. D., Holmes, S. B., Sebastien, R. J., and Rawn, G., *Fenitrothion Risk Assessment*, Canadian Wildlife Service Technical Report Series No. 165, Environment Canada, Ottawa, 1993.

Peakall, D. B., *Animal Biomarkers as Pollution Indicators*, Chapman and Hall, London, 1992.

Peakall, D. B. and L. R. Shugart, Eds., *Biomarkers. Research and Application in the Assessment of Environmental Health*, NATO Advanced Science Institute Series H: Cell Biology, Vol. 68, Springer-Verlag, New York, 1993.

Ratti, J. T. and Scott, J. M., Agricultural impacts on wildlife: problem review and restoration needs, *The Environmental Professional*, 13, 263, 1991.

Rapport, D. J., What constitutes ecosystem health?, *Perspect. Biol. Med.*, 33, 120, 1989.

Suter, G. W. II, A critique of ecosystem health concepts and indexes, *Environ. Toxicol. Chem.*, 12, 1533, 1993.

UNEP, Convention on biological diversity, United Nations Environment Program, Nairobi, Kenya, 1992.

van Monsjou, I. W., Food-fertiliser energy efficiency, in *Fertilizer Society Proceedings*, 152, 1, 1975.

World Commission on Environment and Development, *Our Common Future*, Oxford University Press, Oxford, U.K., 1987.

9 Establishing the Health of Ecosystems

Anne Fairbrother

The application of the health paradigm to entities other than humans or animals holds a great deal of allure because most people have an intuitive sense of what it means to be healthy. However, when questioned more closely, many have difficulty defining exactly what parameters constitute a state of health in themselves, other people, or in animals. This inexactitude is exacerbated when attempting to stretch the definition to ecosystems, which include abiotic components as well as living entities. Therefore, I will argue that the health paradigm is inappropriate as an organizing principle of science and should be applied to ecosystems only as a framework within which diagnostic procedures can be used and that the definition of a healthy state for the environment should remain in the realm of policy.

First, I would like to examine the concepts and definitions embodied in the words "ecosystem" and "health." Ecosystems have been described as self-organizing, open systems (Norton 1992); systems that reduce entropy (Barry Wilson, personal communication 1992); or super-organisms (the Gaia concept) (Lovelock, as referenced in Norton 1992). The Gaia concept, as an embodiment of Clementsian ecology, was rejected by serious intellectuals in the 1960s. The first and second definitions are similar and will be used here. Norton (1992) lists five axioms of ecological management, two of which are particularly relevant to our discussion. These are: the Axiom of Dynamism: nature is more profoundly a set of processes than a collection of objects; all is in flux; The Axiom of Differential Fragility: ecological systems, which form the context of all human activities, vary in the extent to which they can absorb and equilibrate human-caused disruptions in their creative processes.

Keeping these concepts in mind, let us now look at the definition of "health." The dictionary definition of health is (1) the condition of an organism or one of its parts in which it performs its vital functions normally or properly: the state of being sound in body or mind; and (2) flourishing condition (Gove 1966). Health also frequently is defined in the negative as being the absence of disease, although this can become a circular argument. Putting together these definitions of health and ecosystem, a healthy ecosystem can be defined as a system that is active, maintains its organization and autonomy over time, performs its functions normally or properly, and is resilient to stress (Haskell et al. 1992). Note that the concepts of resiliency and sustainability are central to this definition. However, as clearly stated above in Norton's axioms, an ecosystem is a set of processes that are constantly changing. Therefore, no ecosystem is sustainable over evolutionary time. What *is* sustainable,

1-56670-309-3/98/$0.00+$.50

is the trajectory of change of that ecosystem, barring any major cataclysmic forces that substantially redirect the trajectory.

Let us now examine in greater detail the argument that health embodies an agreed upon idealized state. At one end of the spectrum is a "dead" ecosystem; that is, an abiotic system incapable of sustaining life (although we can argue about the definition of "life" should we care to be diverted into environmental existentialism). The other end of the spectrum is much more difficult to define clearly and actually depends upon the social value system of the person making the definition. Some people define a healthy ecosystem as one that will produce big, visible animals that can be caught, photographed, observed, or eaten. Others will define a healthy ecosystem as one that will provide goods and services such as clean water, clean air, board feet of timber, or the ability to absorb pollutants. Still others will define an ecosystem as healthy only if it is in the same successional stage as it was prior to the arrival of nonindigenous peoples. But is a Kentucky bluegrass pasture any less healthy than the deciduous forest that it replaced? The pasture is a self-sustaining ecosystem (at least within a human time frame) and can support a variety of introduced flora and fauna. This debate likely will continue, and it raises the essential issue of whether establishment of a definition of ecosystem health is a scientific or policy question. I submit that the application of the word "healthy" to a predetermined ecosystem state is a matter of policy. As scientists, we need to be involved in such a policy debate, but the elements of the debate go beyond what we write in our scientific journals and include a myriad of social and economic questions as well as fundamental truths concerning the essentiality of an integrative Nature. In other words, the use of the term "ecosystem health" as a definition of an idealized state is not an appropriate paradigm.

But science does support the ecosystem health paradigm if we apply the concepts of health to a process, not an ultimate state. Even here, definitions must be stretched rather thin to apply the paradigm. Moreover, the need for a definition of ecosystem health arises from the requirements of environmental managers and policy makers to set goals against which the consequences of their actions can be judged. Therefore, I suggest that we would be better served to discuss the terms "ecosystem risk characterization," rather than attempting to use the more ambiguous word "health." An ecosystem risk characterization will make either *a priori* or *a posteriori* determinations about the probability of change associated with the imposition of one or more stressors on an ecosystem. Most important, such a characterization will describe the probability of the change of the trajectory along which the ecosystem evolution is naturally occurring (see Figure 9.1). If there are no anthropogenic stresses on an ecosystem, it will change along trajectory 1, where P_1 is the probability of that trajectory being maintained in the face of anthropogenic stress. The probabilities associated with each of the other possible trajectories (P_n) can be determined in a similar manner. While P_n will never equal 1 due to the occurrence of truly random events, sufficient differences among the probability values will exist to enable informed, responsible policy and management decisions to be made.

The converse of this question is to ask whether the past trajectory of change of an ecosystem differs from the trajectory that would have been followed if no anthropogenic stressors had been applied to the system and, if so, which stressor(s) caused

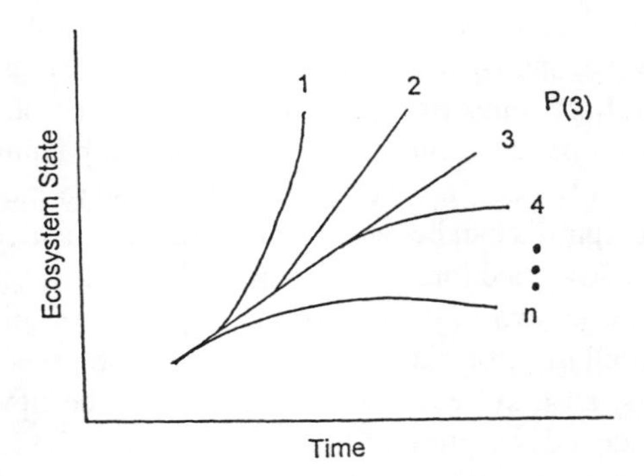

FIGURE 9.1 Theoretical trajectories of ecosystem change. Numbers are arbitrary designations for each trajectory and P(n) indicates the probability of occurrence of the nth trajectory. This figure is a two-dimensional representation of n-dimensional space.

the change. Given that Nature consists of a multiplicity of concurrent and sequential random and/or unpredictable events, there is a large (infinite?) number of possible trajectories that an ecosystem could have taken through evolutionary time, arriving at a current state either similar to or distinctly different from the one observed. However, questions regarding human impact generally are phrased in humanistic time frames (a few hundred years) not evolutionary or ecological time, thereby significantly narrowing the possibilities. The next follow-on question would be to predict how the system would change if the stressors remained or were removed, similar to the *a priori* questions posed in the preceding paragraph. This process of *a posteriori* characterization of ecosystem change and the contribution of various factors to the determination of the trajectory of that change lends itself well to the application of the health paradigm.

There are two biomedical approaches that can be followed. The first is a diagnostic approach that asks: What is wrong with the system? This usually is applied to an individual ecosystem or geographic area. The second approach is an epidemiological one: What are the stressors that have the greatest potential of impacting the system? This requires a population of systems from which to make inferences concerning risk factors. Let us examine these applications of the biomedical approach more closely.

Ecosystem diagnostics for individual systems follows the same deductive logic as is applied to diagnosing the cause of illness in individual organisms. Symptoms are identified that suggest that the system has changed from its previous state. A large number of dead fish in a lake, a reduction in the deer harvest over several years, or chlorotic mottling of vegetation are unusual in most systems and would cause managers to raise questions concerning possible stresses on the system. Once one or more indicators of stress are identified, a history of the area must be compiled. What plant and animal communities were in the area a decade ago? Fifty years earlier? Before agriculture or industry? Is there any information about what kinds of stressors have been applied to the system over the past century or two (both natural stress such as climate changes or volcanic events and anthropogenic stress such as urbanization, agriculture, or industrialization)? This information could be in the form of both oral and written history as well as the interpretation of biotic and geomorphic data (e.g., age of vegetation, spacing of tree rings, pollen deposition in lake cores). General measures of condition should also be obtained at this time, similar to taking the temperature and pulse rate of human patients. Once this background information is available, a description of the system must be formed from which a differential diagnosis can be developed. In human medicine, the system has been well-defined and described for some time (i.e., the physiology and biochemistry of the human body are known in detail) so this step is done prior to the initiation of the diagnostic paradigm. For ecological diagnoses, however, each system will differ from all others, so a system-specific model must be developed. This, at a minimum, must include a description of the major nodes of the food web and major ecological processes of nutrient cycles and energy flows. A differential diagnosis is then prepared, listing all the possible etiologies that may have caused the system to change trajectories. Included in this may be species-specific differential diagnoses (e.g., what caused the fish to die?) as well as a larger system diagnosis (e.g., is the fish die-off an indicator of uncoupling of the phosphorus cycle within the catchment basin?). Continuing to follow the biomedical paradigm, tests, surveys, or analyses would then be conducted to rule in or out each etiology, and a diagnosis would be made. This is always followed by a suggested management strategy (i.e., treatment) that will return the system to a state closer to that which would have occurred if the stress(es) had not been applied. Implicit in this suggestion is a prognostication of the success of the treatment. As with all good medical programs, a follow-up monitoring program should be instituted to ascertain that the treatment is working (in the current jargon, "adaptive management" will be instituted).

Diagnosis of ecosystem dysfunction and monitoring changes of systems through time require the application of methods that measure ecosystem processes and components. These measures can be taken at any level of ecological organization, from molecular to global scales (Figure 9.2). Selection of which level to measure must be driven by the questions being asked. In general, a measurement at one level of ecological organization will allow predictions to be made about that level or the next higher level of organization but will not allow predictions higher in the organizational hierarchy. Conversely, measurements at levels of organization lower than the endpoints of interest will provide valuable information about cause and effect relationships or mechanistic responses that also will be useful for predicting effects in other places or at other times. Additionally, as we move up the scale of ecological organization, we also move to longer temporal scales. Changes happen very quickly at molecular and organ levels, more slowly at organismal or population levels, and even more slowly at the community level which must integrate multiple population changes. Therefore, when selecting methods to assess change in ecosystem condition, we must select the one appropriate to the level of ecological organization, the spatial scale, and the temporal scale of concern.

Frequently, ecosystem managers are interested in changes that occur in humanistic time frames (20 to 100 years). Biomarkers are one way to identify or measure short-term symptoms of stress in indicator species (Peakall and Shugart 1993). Several previous chapters have discussed the application of specific biomarkers to determining ecosystem dysfunctions. All physiological or behavioral changes in

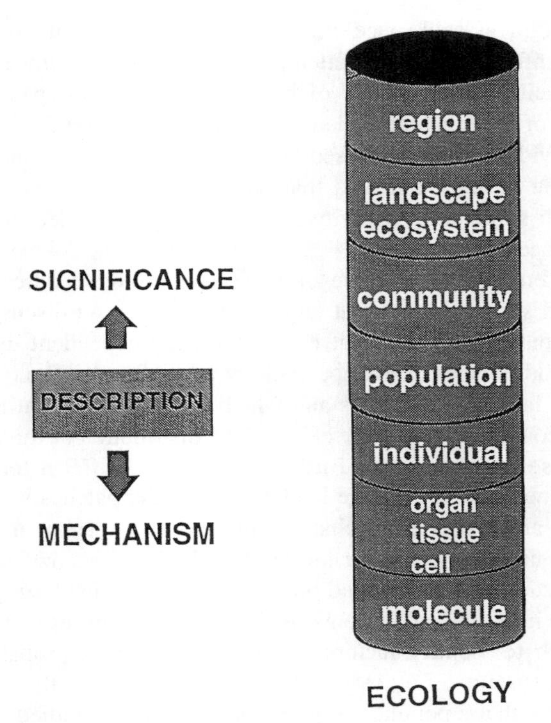

FIGURE 9.2 Representation of the interrelationship of various levels of ecological organization, demonstrating that ecological significance can be extrapolated upward while mechanisms of action are studied at lower levels of organization. (Adapted from Wilson, B.W., 1992.)

animals can result in changes to birth rates, mortality rates, or movements of individuals into or out of an area. These are the three factors that affect population demographics and, therefore, demonstrate how measurements at molecular or organ levels can predict effects at the next higher level of organization, the population (see Figure 9.2). Of course, the question could be raised about why predictions about populations should be made from biomarkers rather than by directly measuring the changes. Certainly it is possible to measure population size or demographics directly, and game managers have been doing just that for the better part of this century. However, measurement of changes in populations requires a longer time and a greater magnitude of stress than do measures of physiological changes, although predictive population models utilizing information from static life tables significantly shorten the time needed for collection of data. A modeling approach also enables a manager to predict with a defined uncertainty the effects of management decisions over long time periods.

Populations of plants or animals are embedded in a food web comprised of many trophic levels that transmit stress impacts throughout the entire community. There have been many attempts to develop a metric of community integrity that would be predictive of system stress. The best known of these is Jim Karr's Index of Biotic Integrity for aquatic systems (IBI) (Karr 1981; Miller et al. 1988). As stated previously by Suter (1993), these types of indices likely are too simplistic because they compress multidimensional space into a single dimension with concomitant loss of a great deal of information. Other attempts at development of measures of community stability include the concepts of keystone species and species richness. The current rhetoric of changing the Endangered Species Act to protect whole ecosystems rather than individual species is based on the assumption that changes in ecological communities can be measured and that these changes occur within a humanistic time frame with sufficient sensitivity so that management decisions can be made before species become extinct.

Lastly, the emerging discipline of landscape ecology addresses questions of biodiversity and system function at large spatial scales. As discussed in an earlier chapter, many patterns observed at large scales are dependent upon processes at lower scales. Additionally, landscape ecology looks for emergent patterns that can only be seen at large spatial scales and which are not dependent upon lower-scale processes. An example of this is the effect of the distribution of forest patches within a landscape on songbird diversity. Bird diversity would differ if forest patches were aggregated in one corner of a large landscape or if the patches were spread evenly throughout the area. Therefore, changes in landscape design may be a stressor independent of alterations at lower levels of ecological organization.

The diagnostic methods touched on above are used most frequently when diagnosing changes in individual systems. A few programs are utilizing the epidemiological approach to identify factors that have the highest probability of causing changes in ecosystems. Following the health paradigm, this epidemiological approach requires that a population of similar systems be studied to correlate environmental risk factors to changes in the system. As with any epidemiological study, this is a correlational approach and additional research may be needed to ascertain whether the correlations are spurious or if they have a true cause and effect relationship. The Environmental Monitoring and Assessment Program (EMAP) of the U.S. EPA is using this epidemiological approach. For example, through a carefully randomized statistical design, a subset of the lakes in the northeastern United States will be monitored for their water quality, fish populations, and related stress factors (e.g., acid precipitation) from which probabilistic statements can be made about the relationship of changes in stress factors to the entire population of lakes in the Northeast. Unfortunately, accurately measuring all possible stress factors has turned out to be problematic, so EMAP likely will only report changes in the populations of ecosystems without attempting to relate the changes to risk factors.

In summary, ecosystems can be defined as self-organizing, open systems comprised of a set of processes. Ecosystem health, defined as a state variable, is a policy issue that goes beyond the boundaries of the natural sciences to encompass the social sciences as well. Therefore, the health paradigm is ill-suited to defining an idealized state for ecosystems, but rather describes a process of logic that can be applied to diagnosing or predicting impacts on ecosystems from a multiplicity of natural or anthropogenic stressors. Lastly, we should remember that the problem of detecting ecosystem disease is one of identifying abnormalities in what is quite naturally a highly irregular pattern and process.

REFERENCES

Gove, P. B., *Webster's Third New International Dictionary of the English Language Unabridged,* G. & C. Merriam Co., Springfield, MA, 1966.

Haskell, B. D., Norton, B. G., and Costanza, R., What is ecosystem health and why should we worry about it?, in *Ecosystem Health: New Goals for Environmental Management,* Costanza, R., Norton, B. G., and Haskell, B. D., Eds., Island Press, Washington, D.C., 1992, Introduction.

Karr, J. R., Assessment of biotic integrity using fish communities, *Fisheries*, 6, 21, 1981.

Miller, D. L., Leonard, P. M., Hughes, R. M., Karr, J. R., Moyle, P. B., Schrader, L. H., Thompson, B. A., Daniels, R. A., Fausch, K. D., Fitzhugh, G. A., Gammon, J. R., Haliwell, D. B., Angermeier, P. L., and Orth, D. J., Regional applications of an index of biotic integrity for use in water resource management, *Fisheries,* 13, 12, 1988.

Norton, B. G., A new paradigm for environmental management, in *Ecosystem Health: New Goals for Environmental Management,* Costanza, R., Norton, B. G., and Haskell, B. D., Eds., Island Press, Washington, D.C., 1992, chap. 2.

Peakall, D. B. and Shugart, L. R., *Biomarkers: Research and Application in the Assessment of Environmental Health,* NATO ASI Series, Series H Cell biology Vol. 68, Springer-Verlag, New York, NY, 1993, 119.

Suter, G. W., A critique of ecosystem health concepts and indexes, *Environ. Toxicol. Chem.*, 12, 1533, 1993.

Wilson, B.W., personal communication, 1992.

10 Establishing the Health of Ecosystems: The Role of Risk Assessment

Thomas E. McKone

ABSTRACT

Risk assessment is a process for identifying adverse consequences and their associated probability. As applied to effects on human and ecological health from environmental contaminants, risk assessment involves four steps: (1) determination of source concentrations or emissions characteristics, (2) exposure assessment, (3) toxicity assessment, and (4) risk characterization. The goal of this chapter is to relate ecological risk assessment to human health risk assessment in order to identify similarities and differences. In the area of public health protection, risk assessment has proven more useful when applied to endpoints with high negative value, such as disasters and irreversible diseases such as cancer and birth defects. Less success has been attained when it is applied to reversible and/or threshold-type diseases such as muscle dysfunction and neurological effects. We will consider why this is and what this implies for ecological risk assessment. An important issue in both health and ecological risk assessment is how precisely we can characterize the distribution among individuals and/or species of potential effects and whether this level of precision favors one or another action.

INTRODUCTION

Risk assessment is a process for identifying adverse consequences and their associated probability. The risk assessment approach differs from other environmental protection strategies in that its principal objective is not to eliminate all risk but to quantify the risk and to balance the level of risk against the cost of risk reduction, against competing risks, or against risks that are generally accepted as trivial or acceptable. Controlling the exposure of ecosystems to environmental contaminants using a risk-based approach requires that we define both an accurate measure for assessing the impacts of contaminants on ecosystems and species and a defensible process for assigning value to the predicted impacts. The purpose of this chapter is to provide an overview of environmental health risk assessment and to define what lessons can be gained from this strategy and transferred to the field of ecological risk assessment. The end-product of a risk-based approach to environmental management is either to identify an acceptable level of exposure or prescribe a required level of technological control.

In order to address the question of whether risk analysis can be applied to the health of ecosystems, we begin with an overview of the risk assessment process — including hazard identification, risk characterization, risk valuation, and risk management. The National Research Council (1982) has divided the practice of risk analysis into two substantially different processes — risk characterization and risk management. In addition to risk characterization and risk management, the risk-based approach begins with a hazard identification and is motivated by risk valuation.

The risk characterization process is considered here in terms of its component steps. Risk characterization is the process of both selecting and quantifying the adverse consequences that result from some action (or inaction) such as the use of a chemical or the use of an industrial process or technology. Risk characterization is used to establish the significance of an estimated risk by defining the magnitude and precision of this estimate. Because of the inherent uncertainty of the risk characterization and risk management processes, it is important to consider how individuals and societies value uncertain adverse consequences. Risk valuation is a component of the risk assessment/management process that is not always explicitly noted. Nevertheless, this component of risk assessment provides the critical link between risk characterization and risk management. The goal of the risk management process is to establish the significance of the estimated risk, compare the costs of reducing this risk to benefits gained, compare the estimated risks to the societal benefits derived from incurring the risk, and carry out the political and institutional process of reducing risk. An important, and often ignored, final step in the risk characterization process is the characterization of uncertainties. This process is so important to the premise behind the risk-based approach and so frequently passed over in actual practice, that the final section of this chapter is devoted just to the issue of confronting uncertainties.

CAN RISK ASSESSMENT BE APPLIED TO THE HEALTH OF ECOSYSTEMS?

Over the last two centuries industrial societies have altered the earth's environment in ways that could have important, long-term ecological, economic, and health implications. Within the fields of geochemistry, ecology, hydrology, and systems analysis, there have been rapid advances in our ability to observe and describe the interactions of the earth system at many scales. Environmental-health-risk and ecological-risk assessments are cross-disciplinary approaches that have evolved in part to address these types of potentially detrimental impacts. Risk analysis involves interfaces among the fields of systems science, operations research, statistics and probability, welfare economics, public health, biology, and earth sciences, just to name a few.

In addition to risk characterization and risk management, the risk-based approach begins with a hazard identification and is often motivated by risk valuation. Without the identification or perception of a hazardous condition, there will be little or no effort to further characterize and manage risk. The goal of risk characterization is to develop models and/or measurements that can be used to determine the magnitude of risk, parameters that contribute to this magnitude, and the likely uncertainty about this magnitude. Risk management is the process of weighing policy alternatives and selecting the appropriate societal or institutional response. This last step is used to integrate the results of a risk characterization with social, economic, and political valuation to reach a decision. Linking these processes is the concurrent effort to evaluate risk. Risk evaluation is the process by which the risk characterization and risk management processes are reconciled with individual and societal valuations of risk. Figure 10.1 provides a view of how the risk-analysis process might proceed for assessing the health and/or ecological impacts of a toxic chemical. Each of the major steps in this process involves one or more actions that are listed to the right of each major step.

Hazard Identification

It is typical for a risk assessment to begin with efforts to identify the potential hazards associated with a chemical and its use. The distinction between risk and hazard can be illustrated by the example of making the choice to cross the ocean in either a rowboat or a cruise ship. Both modes of transit provide a similar hazard (drowning), but the risk associated with this hazard differs substantially between the two modes of transit.

In the human health risk field, the health hazards of interest involve cancer, reproductive effects, irreversible chronic effects, reversible chronic impacts, irreversible acute effects, and reversible acute effects. All of these are considered unacceptable events when the cause-effect pathway is certain. When the cause-effect pathway is uncertain, we work to make the risk of these impacts diminishingly small. In ecological risk assessment the hazard states of concern have not been so clearly delineated. There is a need to ask whether it is the hazard to individual animals, to a given population, to a species, or to an ecosystem that defines an obvious hazard state.

Human health hazard assessment is based on either human epidemiological studies, animal toxicity studies, or *in vivo* cell culture studies. These are combined with simple measures of potential exposures such as environmental concentrations or source term estimates to provide a more complete picture of hazard.

Risk Characterization

For a substance that has been found to pose a significant hazard, a regulator following a risk-based strategy will carry out a risk characterization study in order to select risk-management options. As applied to a toxic chemical, risk characterization includes five principal elements: (1) quantification of releases and environmental concentrations in the vicinity of a source; (2) quantification of exposure and doses to the target population and how the dose is distributed among the population; (3) quantification of a dose-response function for all potential toxic hazards that have been identified; (4) estimates of the numbers and severity of consequences expected

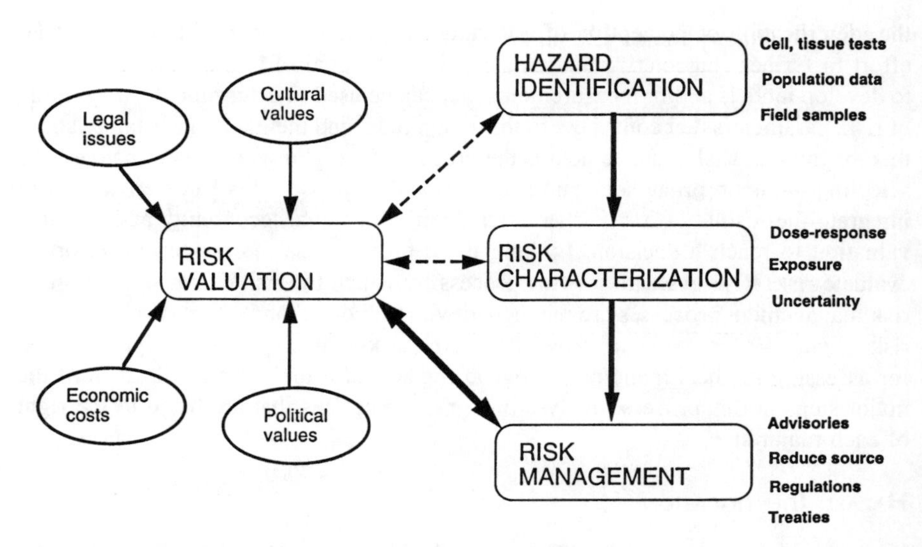

FIGURE 10.1 Schematic diagram of the risk assessment/management process as it is applied to human and ecological health. This diagram emphasizes the importance of hazard and risk valuation in affecting all the stages of the risk-assessment/management process.

within the population at risk; and (5) an assessment of the magnitude and sources of uncertainty that limit the precision of the estimate of consequences.

Dose-Response Assessment

Detrimental effects of toxic chemicals on humans, animals, and plants can be classified as "stochastic" or "nonstochastic" effects (ICRP 1977). Stochastic effects are those for which the probability of an effect occurring, rather than the severity of the effect, is proportional to dose, without threshold. Nonstochastic effects are those for which the severity of effect is a function of dose and for which a threshold may exist. For example, the human health effects of carcinogens and many types of genetic effects are assumed to be stochastic. In contrast, the effects of neurotoxins, such as lead and mercury, are assumed to be nonstochastic. Also, effects of an essential element such as selenium are nonstochastic with an evident, but often unknown, threshold.

Among the many issues that complicate the process of establishing a dose-response function is susceptibility. In large heterogeneous populations there are large variations in susceptibility to toxic effects, due in part to variations in genetic predisposition to certain disease states, age, physical stresses, and other chemical exposures that may be extant in the system of interest. It is almost impossible to design laboratory or field studies that will capture the impact that such susceptibility variations might have on a single or multiple animal (plant) species. Variations in susceptibility are an issue for both stochastic and nonstochastic effects.

Another problem associated with the common use of simple linear dose-response models is the failure to recognize the robustness displayed by species and ecosys-

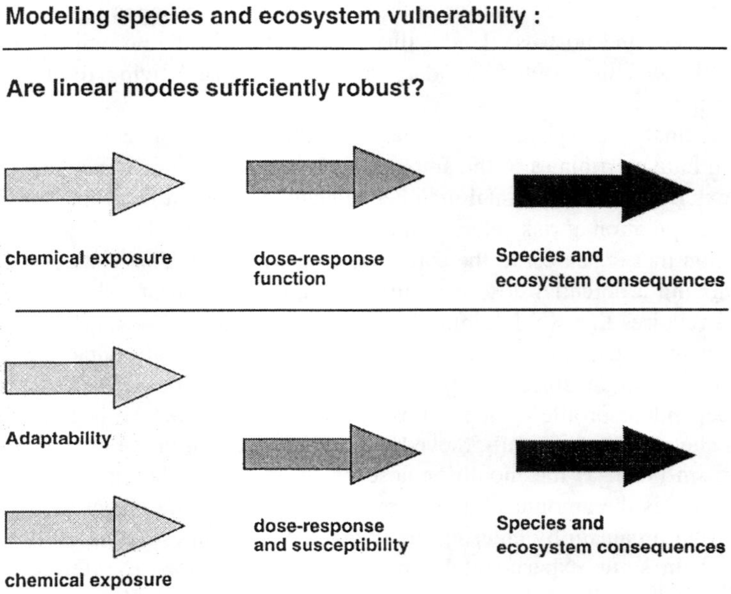

Modeling species and ecosystem vulnerability :

Are linear modes sufficiently robust?

chemical exposure

dose-response function

Species and ecosystem consequences

Adaptability

chemical exposure

dose-response and susceptibility

Species and ecosystem consequences

FIGURE 10.2 The difference between the linear model relating chemical exposure and a dose-response function to consequences and a nonlinear model that addresses the observation that species (including humans), social systems, and ecosystems can adapt to some level of stress while showing few consequences but then display a precipitous dose-response function when a certain threshold for multiple stresses is exceeded.

tems. In this context robustness refers to the resistance to significant size fluctuations displayed by a species population or an ecosystem. The robustness reflects adaptability. A linear model relating chemical exposure and a dose-response function to consequences fails to capture the type of adaptation observed in a robust system over a rather large range of chemical concentrations. However, when multiple stresses increase to the point where the population can no longer adapt to stress, then the population will display a much larger dose-response function. In this situation a nonlinear model or a second-order dose-response function is needed. This particular problem arises in ecological risk for both stochastic and nonstochastic effects. This type of situation is illustrated in Figure 10.2.

Exposure and Dose Assessment

Exposure assessments contribute to a number of health and environmental assessments, including risk assessments, status and trends analyses, and epidemiological studies. Previously, exposure was defined by the U.S. EPA (1988) in terms of contact with the so-called "exchange boundaries" of an organism where contaminant absorption takes place (skin, plant surfaces, lung, gills, gastrointestinal tract). However, the more recent consensus of the scientific community (National Research Council, 1991a, 1991b; U.S. EPA, 1992) suggests that exposure should be defined in terms of contact with the visible exterior of the organism (skin and openings into the body, such as mouth and nostrils). Under this definition, we view an animal or plant as having a hypothetical outer boundary separating internal living tissues from the outside surfaces.

In risk analysis, exposure assessments are used to translate contaminant sources into quantitative estimates of the amount of chemical that comes in contact with the visible exterior of an individual member selected at random from the population at risk. The population at risk refers to the species that does or plausibly could inhabit the location that is nearest to the source of contamination. This contact is the basis for estimating a potential dose used in the characterization of risk. An assessment of intake requires that we determine how much crosses the exchange boundaries of the organism. Thus, exposure is the condition of a chemical contacting the outer boundary of an organism, and exposure over a period of time can be represented by a time-dependent profile of the exposure concentration. *Intake* is the process by which a chemical is physically moved through an opening in the outer boundary of an organism (such as the mouth or nose of an animal or the stomata of a plant). *Applied dose* is the amount of a contaminant that comes in contact with the living tissues of an organism by entering the lungs, the gastrointestinal tract, or the outer skin layer. In some experimental designs, the applied dose is referred to as the administered dose. *Potential dose* is an approximation to applied dose that is simply the amount of chemical in the food or water ingested, air inhaled, or material applied to skin surface. *Absorbed dose* is the amount of contaminant penetrating the exchange boundaries of an organism after contact. Absorbed dose is calculated from the intake and absorption efficiency. For animals it is typically expressed as the mass of contaminant absorbed into the body per unit mass per unit time such as mg/kg-d.

Defining exposure pathways is an important component of the exposure assessment. An exposure pathway is the course a chemical or physical agent takes from a source to an exposed organism. An exposure pathway describes a unique mechanism by which an individual or population is exposed to chemicals or physical agents at or originating from a source. Each exposure pathway includes a source or release from a source, an exposure point, and an exposure route. If the exposure point differs from the source, a transport/exposure medium (such as air) or media (in cases of intermedia transport, such as water to air) also is included.

Exposure assessments often rely implicitly on the assumption that exposure can be linked by simple parameters to ambient concentrations in air, water, and soil. However, for human populations, total exposure assessments that include time and activity patterns and micro-environmental data reveal that an exposure assessment is most valuable when it provides a comprehensive view of exposure pathways and identifies major sources of uncertainty. Thus, we see the need to address many types of "multiples" in the quantification of exposure and dose, such as multiple media (air, water, soil), multiple exposure pathways (or scenarios), multiple routes (inhalation, ingestion, dermal), and multiple target tissues for dose and effect.

Exposure model guidelines have been described by the U.S. Environmental Protection Agency (U.S. EPA, 1989, 1992) and the State of California Department of Toxic Substances Control (DTSC, 1992a, 1992b).

Combining Exposure and Dose-Response Estimates to Characterize Risk

Once exposure information and a dose-response model have been generated, risk characterization is carried out by constructing a model for the distribution of individual lifetime risk, H(t), at some time t in the future within a population exposed for an exposure duration, ED (years), to a quantity S_m (in mg) of contaminant introduced to compartment *m* at time zero. This is done theoretically by summing the dose and effect over exposure routes, over environmental media, and over exposure pathways.

$$H(t) = S_m(0) \times \left\{ \sum_{\substack{\text{j routes,}}} \sum_{\substack{\text{k environmental} \\ \text{media,}}} \sum_{\substack{\text{i exposure} \\ \text{media}}} \left[Q_j\left(ADD_{ijk}\right) \times \left(\frac{ADD_{ijk}}{C_k}\right) \times \Phi\left[S_m(0) \rightarrow C_k, t\right] \right] \right\} \quad (1)$$

where $\Phi[S_m(0) \rightarrow C_{k,}t]$ is the multimedia dispersion function that converts the contaminant release $S_m(0)$ mg measured today, into contaminant concentration C_k at a time t in the future for a duration ED in environmental medium k (units of C_k are mg/kg for soil, mg/m^3 for air, and mg/L for water). (ADD_{ijk}/C_k) is the unit dose factor, which is the average daily potential dose (over a specified averaging time) from exposure medium i by route j (inhalation, ingestion, dermal uptake, etc.) attributable to environmental compartment k divided by C_k when C_k is constant over the duration ED. The exposure media summation is over the number of exposure media that link potential dose by route *j* to contaminants in compartment k. $Q_j(ADD_{ijk})$ is the dose-response function that relates the potential dose, ADD_{ijk}, by route *j* to the lifetime probability of detriment per individual within the population.

Risk Valuation

Risk valuation is a component of the risk assessment/management process that is not always explicitly noted. Nevertheless, this component provides the critical link between risk characterization and risk management. Also, in many situations there are subtle links between the valuation of risk and the choice of what hazards to assess and what risks to characterize (see Figure 10.1).

The decision to expend societal resources to identify, estimate, and manage risk carries with it an implicit valuation of the risk being controlled. Because of the inherent uncertainty of the risk characterization and risk management processes, it is important to consider how individuals and societies value uncertain adverse consequences. We expect such valuations to be expressed in terms of relative preferences, economic preferences, or ethical constraints.

In the valuation phase of the risk-based approach, the results of a risk characterization are integrated with social, economic, and political considerations to provide input to the risk-management process. A variety of techniques have been proposed and used to systematically apply the values held by different stake-holders to the evaluation of risks. Some of the commonly used techniques of risk valuation include the elicitation of individual and societal preferences, decision analysis, theories of science policy, social-welfare economics, and ethics. One or more of these valuation strategies can provide an important input to the risk management decision.

Even though exposure and dose-response estimates are an important part of understanding risks, they fail to provide the complete picture on how individuals and societies decide to manage risk. It is also important to understand the process by which people perceive health and ecological risks and then decide how bad these risks are. Within the field of experimental psychology, there has been a recent examination of the processes by which people perceive and compare risks and how this affects risk management. Kahneman et al. (1982) have prepared a review of this work. The economic problem of individual and societal valuation of life has been considered by Raiffa et al. (1977). Since human life span is not a marketable commodity, it is difficult to define a value of a statistical year of life lost. Many ecological resources have this same characteristic. That is, because they are not industrial or consumer commodities, simple market values fail to establish the societal value of ecological resources.

Because risk assessment is used when a decision must be made in the absence of complete information, it is a process that implicitly involves a judgment made with much uncertainty. Psychologists have observed that when people are asked to make judgments involving uncertainty, they subconsciously adopt a number of heuristics or rules of thumb for decision making. In particular, their belief about the likelihood or severity of an event is related to ease with which they can recall previous occurrences of the given event or a similar event. Kahneman et al. (1982) reveal a number of interesting rules regarding the acceptability or trading of risks, among them: people have a preference for reducing one risk to zero instead of lowering multiple risks; gains and losses are not valued the same; also, people are not willing to trade gains against losses; people are not willing to make risk-risk trade offs; and (with the exception of weathermen and odds makers) most experts overestimate the precision of their predictions of any adverse event.

Risk Management

Once the risks associated with an action have been identified and quantified, it becomes necessary to develop a basis for evaluating these risks and then, if necessary, developing and carrying out some actions to reduce the risks. There are four types of analyses that are commonly used in the risk management process — risk-benefit, cost-benefit, risk-risk, and cost-effectiveness analyses. A *risk-benefit analysis* provides a comparison of the risks added by an activity to the concurrent benefits (usually economic) provided to society. A *cost-benefit analysis* relates the financial cost (in dollars) of reducing risk to the benefits (in equivalent dollars or an appropriate surrogate) gained by reducing risk. A *risk-risk analysis* establishes the significance of an estimated risk by comparing it to some other acceptable or familiar risk such as those associated with background exposures, household accidents, occupational hazards, natural occurrences, etc. A *cost-effectiveness analysis* is used to compare risk reduction per unit cost among several options for dealing with the same risk.

CONFRONTING UNCERTAINTIES

Regulatory toxicology and risk assessment often operate under the premise that, with sufficient funding, science and technology will provide an obvious and cost-effective solution to the problems of protecting human health and the environment. However, in reality there are many sources of uncertainty and variability in the process of human-health and ecological risk assessment. Many of these uncertainties and variabilities are not reducible. Effective policies are possible under conditions of uncertainty, but such policies must take the uncertainty into account. There is a well-developed theory of decision making under uncertainty, which is described in Chernoff and Moses (1959), Lindley (1985), and Berger (1985) among others. One often-used method for addressing uncertainty in risk assessments is the compounding of upper bound estimates in order to make decisions based on a highly conservative estimate of exposure and risk. Such an approach is contrary to the principles of decision making under uncertainty (as described in the texts cited above). This latter approach leaves the decision maker with no flexibility to address margins of error; to consider reducible vs. irreducible uncertainty; to separate individual variability from true scientific uncertainty; or to consider benefits, costs, and comparable risks in the decision-making process. Figure 10.3 provides an illustration of how the three component steps of the risk assessment process relate to science, uncertainty, and policy. The three barometers next to these steps reveal the relative concentration of science, uncertainty, and policy. The darker shade indicates a higher concentration. We see that the greatest concentration of the scientific method is in the hazard-identification phase. As we move to the characterization and management of risk, we must make use of more and more uncertain information. The use of uncertain information requires more policy input.

The principles of decision making under uncertainty are not necessarily complex. Often the principles of such decision making are simply common sense. But in any issue involving uncertainty, it is important to consider a variety of plausible hypotheses about the world; consider a variety of possible strategies for meeting our goals; favor actions that are robust to uncertainties; favor actions that are informative; probe and experiment; monitor results; update assessments and modify policy accordingly and favor actions that are reversible (Ludwig et al. 1993).

In order to make an ecological risk assessment consistent with such an approach, it should have both sensitivity and uncertainty analyses incorporated directly into the entire process of risk characterization.

Uncertainty and Sensitivity Analysis

Rarely can we measure the magnitude of chemical exposure and the resulting ecological risk. Thus, such outcomes must be estimated by models. Chemical fate, exposure, and ecological effects models range from simple "rule-of-thumb" models to complex stochastic models. The reliability of these models is determined by the

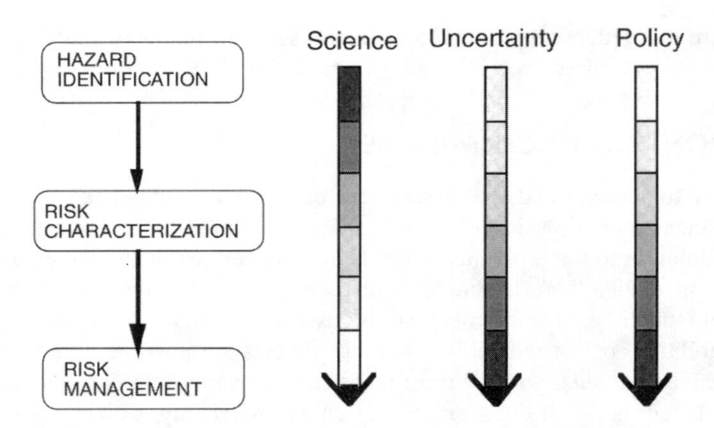

FIGURE 10.3 An illustration of how the three component steps of the risk assessment process relate to science, uncertainty, and policy. The three barometers next to these steps reveal the concentration of science, uncertainty, and policy. The darker shade indicates a higher concentration.

precision of the model inputs and the accuracy with which the model addresses the relevant physical, chemical, and biological processes. There are two approaches by which we can assess how model predictions are impacted by model reliability and data precision — uncertainty analysis and sensitivity analysis. In order to address sensitivity and uncertainty, one can think of a model as producing an output Y, such as risk, that is a function of several variables, X_i, and time, t,

$$Y = f(X_1, X_2, X_3, \ldots X_k, t). \tag{2}$$

The variables, X_i, represent the various inputs to the model. As applied to mathematical models, uncertainty analysis involves the determination of the variation or imprecision in the output function based on the collective variation of the model inputs, whereas sensitivity analysis involves the determination of the changes in model response as a result of changes in individual model parameters.

Uncertainty Analysis

Describing uncertainty in the output variable, Y, involves quantification of the range of Y, its arithmetic mean value, the arithmetic or geometric standard deviation of Y, and upper and lower quantile values of Y, such as 5% lower bound and 95% upper bound. Convenient tools for presenting such information are the *probability density function* (PDF) or the *cumulative distribution function* (CDF) for Y. However, the PDF or CDF of Y can often only be obtained when we have meaningful estimates of the probability distributions of the input variables X_i.

Sensitivity Analysis

The goal of a sensitivity analysis is to rank the input parameters on the basis of their contribution to variance in the output. Sensitivity analyses can be either global or local. A *global sensitivity analysis* quantifies the effects of variation in parameters over their entire range of values. A global sensitivity analysis requires an uncertainty analysis as a starting point. The variance in the outcome is compared to the variance of the inputs. A *local sensitivity analysis* is used to examine the effects of small changes in parameter values at some defined point in the range of outcome values.

Uncertainty and Variability

One of the issues in uncertainty analysis that must be confronted is how to distinguish between the relative contribution of true uncertainty vs. inter-individual variability (i.e., heterogeneity) to the characterization of predicted population risk (Bogen and Spear 1987). Uncertainty or model-specification error (e.g., statistical estimation error) can be modeled using a random variable with an identified probability distribution. In contrast, inter-individual variability refers to quantities that are distributed empirically within a defined population. Such factors are food ingestion rates, exposure duration, and expected lifetime. Variability and true uncertainty have also been referred to as, respectively, *Type A* uncertainty, that "due to stochastic variability with respect to the reference unit of the assessment question," and *Type B* uncertainty, that "due to lack of knowledge about items that are invariant with respect to the reference unit of the assessment question" (IAEA 1989). There are situations in which true (Type B) uncertainty is negligible relative to variability (Type A) uncertainty, and in these situations, the outcome of a variance propagation analysis simply represents the expected statistical variation in dose or risk among the exposed population. When neither variability nor uncertainty is negligible, we have a situation in which there are multiple CDFs representing variability, but the correct curve is unknown because of uncertainties.

A Tiered Approach to Uncertainty Analysis

In order to more directly confront uncertainties in risk assessments, it is necessary to take a tiered approach to uncertainty analysis. Three tiers are needed.

First, the variance of all input values should be clearly stated, and the impact of these variances on the final estimates of risk assessed. At a minimum, this can be done by listing the estimation error or the experimental variance associated with the parameters when these values or their estimation equations are defined. It would help to define and reduce uncertainties if a clear summary and justification of the assumptions used for each aspect of a model are provided. In addition, it should be stated whether these assumptions are likely to result in representative values or conservative (upper bound) estimates.

Second, a sensitivity analysis should be used to assess how model predictions are impacted by model reliability and data precision. The goal of a sensitivity analysis is to rank the input parameters on the basis of their contribution to variance in the output.

Third, variance propagation methods (including but not necessarily Monte-Carlo methods) should be used to carefully map how the overall precision of risk estimates is tied to the variability and uncertainty associated with the models, inputs, and scenarios.

SUMMARY AND CONCLUSIONS

There is growing concern that human activities, principally those related to industrialization, are producing sustained ecological stresses that may lead to severe adverse effects on a regional scale. Although the precise degree of the changes and the exact magnitude of the consequences globally and regionally remain uncertain, the implications are that regionally the resilience of whole ecosystems could be impacted. Consequently, with an absence of complete information, decision makers at a regional level are confronted with the task of formulating defensible, proactive strategies that will both avert substantial environmental loss and limit the burden of unnecessary costs for their society. Accordingly, to construct an effective strategy a decision maker must first interpret the significance of uncertainties (both qualitative and quantitative) in the scientific predictions linking ecological effects to chemical releases and then balance these predictions against an uncertain regional economic forecast driven both by the global economy and by human behavior. Such a strategy can be based on risk assessment, which is a field that has evolved to deal with the problem of addressing uncertain adverse consequences.

As applied to environmental contaminants, risk assessment involves four steps: (1) determination of source concentrations or emissions characteristics, (2) exposure assessment, (3) toxicity assessment, and (4) risk characterization. This chapter summarizes a strategy based on risk assessment for evaluating the sources of uncertainty in predictive ecological-risk assessments. The limitations of the risk estimation methods presented here make clear that risk managers should be aware of the uncertainty in risk estimates and include this awareness in their decisions and their communications of risk to the public. When uncertainties are large, the values of the stake holders are as important as the risk estimates. Furthermore, this situation suggests the need to consider carefully the uncertainties of model assumptions and inputs so that effort is directed at those components having the largest contribution to overall variance in model predictions.

ACKNOWLEDGMENTS

This work was performed under the auspices of the U.S. Department of Energy (DOE) through Lawrence Livermore National Laboratory (LLNL) under Contract W-7405-Eng-48.

REFERENCES

Berger, J.O., *Statistical Decision Theory and Bayesian Analysis*, Springer-Verlag, New York, 1985.

Bogen, K.T. and Spear, R.C., Integrating uncertainty and interindividual variability in environmental risk assessments, *Risk Analysis,* 7, 427, 1987.

Chernoff, H. and Moses, L.E., *Elementary Decision Theory*, Wiley, New York, 1959; reprinted by Dover, New York, 1986.

Department of Toxic Substances Control (DTSC), Documentation of assumptions used in the decision to include and exclude exposure pathways, in *Guidance for Site Characterization and Multimedia Risk Assessment for Hazardous Substances Release Sites*, Volume 2, Report UCRL-CR- 103462, Lawrence Livermore National Laboratory for State of California, Department of Toxic Substances Control, Livermore, CA, 1992a, chap 2.

Department of Toxic Substances Control (DTSC), Guidelines for the documentation of methodologies, justification, input, assumptions, limitations, and output for exposure Models, in *Guidance for Site Characterization and Multimedia Risk Assessment for Hazardous Substances Release Sites*, Volume 2, Report UCRL-CR-103460, Lawrence Livermore National Laboratory for State of California, Department of Toxic Substances Control, Livermore, CA, 1992b, chap. 3.

International Atomic Energy Agency (IAEA), *Evaluating the Reliability of Predictions Made Using Environmental Transport Models*, Safety Series 100, International Atomic Energy Agency, Vienna, 1989.

International Commission on Radiological Protection (ICRP), *Recommendations of the International Commission on Radiological Protection*, ICRP Report No. 26, Pergamon Press, New York, 1977.

Kahneman, P., Slovic, P., and Tversky, A., *Judgment Under Uncertainty: Heuristics and Biases*, Cambridge University Press, New York, 1982.

Lindley, D.V, *Making Decisions*, 2d Ed., Wiley, New York, 1985.

Ludwig, D., Hilborn, R., and Walters, C., Uncertainty, resource exploitation, and conservation: lessons learned from history, *Science*, 260, 17, April, 1993.

National Research Council, *Risk and Decision Making: Perspectives and Research,* National Academy Press, Washington, D.C., 1982.

National Research Council, *Frontiers in Assessing Human Exposure to Environmental Toxicants,* National Academy Press, Washington, D.C., 1991a.

National Research Council, *Human Exposure Assessment for Airborne Pollutants: Advances and Opportunities*, National Academy Press, Washington, D.C., 1991b.

Raiffa, H., Schwartz, W.B., and Weinstein, M.C., Evaluating health effects of societal decisions, in *Decision Making in the Environmental Protection Agency*, Volume II, Committee on Environmental Decision Making, Commission on Natural Resources, National Academy of Sciences, National Academy Press, Washington, D.C., 1977.

U.S. Environmental Protection Agency (U.S. EPA), Proposed guidelines for exposure-related measurements, *Federal Register*, 53, 48830, December 2, 1988.

U.S. Environmental Protection Agency (U.S. EPA), *Risk Assessment Guidance for Superfund Volume I Human Health Evaluation Manual (Part A)*, Report EPA/540/1-89/002, Office of Emergency and Remedial Response, 1989.

U.S. Environmental Protection Agency (U.S. EPA), Guidelines for exposure assessment: notice, Environmental Protection Agency, *Federal Register*, 57(104), 22888, May 29, 1992.

11 Stressors in the Marine Environment

Edward D. Goldberg

INTRODUCTION

Over the past fifty years, marine stressors (or pollutants, substances introduced to the environment which result in resource loss, be it public health, ecosystem integrity, or aesthetics) have been identified both by catastrophes and by scientific insights. For example, in the 1950s and 1960s the entry of mercury from a chemical plant on the shores of Minimata Bay, Japan, into the marine food chain caused consuming citizens to come down with neurological illnesses and even deaths. The culprit, methyl mercury, was identified in 1970. Perhaps the most toxic substance ever deliberately introduced into the ocean system, tributyl tin, was the cause of a near-wipeout of the oyster industry in the Bay of Arcachon, France. The chemical is used as an antifouling agent in marine paints. It leaches from the hulls of vessels into seawater where it can affect nontarget organisms, such as oysters, algae, and gastropods. The problem came to light in the 1980s, and tight restrictions upon its use have mitigated its impact. The classic pollution problem involving DDT, whose destruction of nontarget organisms, especially fish-eating birds, was first identified in the 1950s by U.S. Fish and Wildlife investigators and was later scientifically and poetically described by Rachel Carsons in her 1962 volume *Silent Spring*. In 1972 the U.S. EPA brought DDT usage under severe regulation. This event signaled unacceptable ecosystem alteration as a criterion, in addition to that of public health, in the control of toxic chemical entry into the environment.

Of equal importance in stressor regulation has been the introspection of scientists. In the early 1950s scientists in the U.S., U.K., U.S.S.R., and France, among other nations, recognized that the promiscuous release of artificially produced radionuclides from power plants and weapons testing to the environment could endanger public health through the consumption of foods from the sea or through exposure. As a consequence of both national and international deliberations, guidelines for the controlled entry of these toxic materials were formulated, and the oceans and atmospheres did not receive amounts that have endangered societal health.

THE SEARCH FOR STRESSORS

Recent advances in environmental sciences have led to the identification of potential pollutants. For example, developments in the understanding of the relative toxicities

1-56670-309-3/98/$0.00+$.50

of polychlorinated biphenyls can lead to more focused monitoring tactics. The coplanar members of this group pose far more serious problems than their non-coplanar counterparts. Or again, episodes with metals suggest only three have reduced marine environmental quality with respect to public and/or ecosystem health: copper, tin, and mercury. Copper released from reprocessing plants in Taiwan was accumulated by maricultured oysters and led to their deaths. There are also the previously mentioned examples of methyl mercury and tributyl tin. These pollution episodes have certain commonalities: the forms of the metals were organic complexes (seawater copper is presumably associated completely with organic ligands); the impacts upon life processes were noted in maricultural activities; and catastrophes were necessary to bring the pollutants to the attention of the scientific community.

The formation of novel strategies to diagnose the health of the oceans evolves from these past incidents. They might include:

1. Systematic and routine surveillance of maricultural activities for any unusual morbidities or mortalities.
2. Surveys of metals and metalloids which are complexed with organic ligands in coastal waters.
3. Systematic determinations of metal entries to coastal waters from industrial, agricultural, and social activities.

Clearly, any such activities, if carried out, would place an additional burden upon the already limited resources of agencies responsible for understanding the health of the coastal ocean. I submit that some of the monitoring activities carried out by such organizations as NOAA Status and Trends, the California Mussel Watch, and EPA's EMAPS are far too elaborate and do little to give an effective evaluation of the quality of coastal environments. Judicious elimination of ineffective programs would free up resources to carry out innovative and cost-effective activities.

A dedicated scientific community also seeks the identification of potential marine stressors. Often brought together by international agencies, periodic meetings assemble conventional wisdom from many countries. The last large affair, orchestrated by the United Nations Environment Program in 1990 on "The State of the Marine Environment," pinpointed potential marine environmental problems. From their listing I will consider three, which at the present time are receiving, in my opinion, inadequate attention: the entry of plant nutrients, such as nitrate and phosphate, which encourage eutrophication; plastics on the coastal sea floor; and artificial radionuclides. My criteria for inclusion were: substances that have long residence times in the oceans; substances that have constant to increasing fluxes to the coastal zone; and substances that act as real pollutants, i.e., can impact coastal resources.

NUTRIENTS, PLASTICS, AND RADIONUCLIDES

Perhaps the most important of the understudied potential stressors encompass the plant nutrients which form the base of the eutrophication problem, well recognized by national and international agencies but inadequately addressed. Few systematic studies are being made upon vulnerable areas with respect to the levels of bionutrients and their changes with time and the make-up of ecosystems, especially the populations of plankton. Further, the quantification of areas of hypoxia and anoxia in waters and sediments is poorly studied as well as other potential effects linked to eutrophication such as toxic algal blooms, red tides, etc. I submit that it is a national disgrace in spite of deliberations by many scientific groups in the past that we have no coordinated U.S. program.

High priority survey sites, drawn from areas with large urban populations and/or agricultural, have been identified by a committee of the Marine Board of the U.S. National Research Council and include: New York Bight/Long Island Sound, Chesapeake Bay, and the northern Gulf of Mexico. Because of uncertainties in the controlling factors of eutrophication, a large set of measurements has been proposed. Further, it was evident that large areas and long time periods (decades) would be required for the program.

The accumulation of plastic debris on the coastal sea floor poses a potential jeopardy to the benthic community through its interference with the exchange of dissolved gases between overlying waters and pore waters. The framers of the International Convention for the Prevention of Pollution from Ships (MARPOL) through its ANNEX V were most far-sighted in proposing a total prohibition on the discharge of plastic materials. There are many observable destructive actions of plastic wastes. They can have unacceptable impacts upon living organisms through ingestion or entanglement. Interferences with ship operations through fouling of propellers, water intake pipes, and fishing gear have been costly to remedy. On the beaches and in surface waters, they offer aesthetic insults.

But perhaps the most pressing potential danger lies in their accumulation on the sea floor. Although freshly introduced plastics are buoyant, they quickly accumulate organic coatings which sorb shells, sand, and other debris and sink to the bottom with increased densities. The plastic artifacts provide habitat for opportunistic organisms. But their inhibition of gas exchange processes can lead to anoxia and hypoxia in the deposits. These effects may seriously interfere with the normal functioning of ecosystems and may alter the make-up of life on the sea floor.

Innovative monitoring programs to ascertain whether or not the coverage of the coastal sea floor by plastics is increasing should be initiated in high population areas. A variety of options for surveys are available: photographic, diving, submersible, or trawl. Trawl surveys appear to be the least expensive and perhaps can provide the most statistically satisfying results which can then be used as a springboard for ecological studies.

A final stressor problem involves the past and present husbandry of artificial radionuclides. It has been a decade since they have been a part of U.S. monitoring programs. The last time that I am aware of involved the U.S. National Mussel Watch Program from 1976 to 1978. During that time the sources of the radionuclides were primarily wastes from nuclear energy facilities with minor contributions from weapons testing. Still, long-lived radionuclides continue to enter and to accumulate in the marine environment.

But a new set of concerns has developed — entries from the transocean shipments of processed fuels and fuels awaiting to be processed and the disposal of nuclear submarine reactors and radioactive wastes. In the first instance, the plutonium carrier *Ataksuki Maru* in late 1992 sailed from Cherbourg, France, to Japan with a cargo of 1.5 tons of plutonium oxide, enough plutonium to construct 120 nuclear bombs. The possibility of loss from fire or explosion enroute is augmented by a dispersion from terrorist activity.

Reports from the former Soviet Union describe the dumping of thirteen submarine reactors and 17,000 barrels of radioactive waste into the coastal waters of the Kara Sea between 1968 and 1982. With time, the reactor hardware may lose integrity through corrosion with a consequential loss of radioactivity. Clearly, continuous monitoring is called for where the disposed materials are not recovered. The monitoring of artificial radionuclides in the environment might be accomplished by "Mussel Watch" strategies in appropriate locations such as the northern seas and in the Mediterranean and North Seas where large-scale reprocessing of nuclear wastes takes place.

THE PROBLEM OF MULTIPLE STRESSORS

The cause-effect relationship between a single stressor and an impact has been established in the marine environment in only a few cases. The mobilization of calcium by DDT and its degradation products has been invoked to explain egg-shell thinning and the population fall-off among marine birds. The neurotoxicity and hair and/or blood burdens of methyl mercury have been successfully related in the Minimata Bay incident. The incidence of imposex, a disease in which females acquire male sexual characteristics with subsequent population failures, has been shown to follow the levels of tributyl tin in gastropods.

What about multiple pollutant impact upon marine organisms? This may be a main ecotoxicological problem of the 1990s. Establishing biochemical stresses that lead to morbidities or mortalities is a rather tall order. I suspect that entry to the general problem will be through population declines, hopefully not catastrophic. Suspicions abound that the global declines in seal populations may arise from the effects of multiple pollutants. It must be emphasized that nonchemical stresses, such as temperature or crowding, can couple with chemical stresses to bring about malfunctioning of marine ecosystems.

12 An Attempt to Explain Ecological Health with a Metaphor

Bill L. Lasley

I want to use human reproductive health and reproductive epidemiology as metaphors for ecological health and the discipline of studying ecological health. I will first use reproductive health in a human population as an example of a concept that has different meanings to different people... much like ecological health has different meanings to many scientists. I then want to demonstrate how human reproduction is managed. It is managed in many different ways to achieve very different outcomes depending upon local customs and/or economies much like many of our ecosystems. Finally, I will use the metaphors to illustrate that sometimes the same process or event can have a very different economic or emotional impact on different people.

Immunoassays of one sort or another are critical tools to the fertility specialist of today, and I will predict they will be central to future work in many other disciplines. Like a small ecosystem, the reproductive system is governed by dynamic changes in many different but interacting components. Immunoassays are critical in monitoring these changes, and the reproductive status of women and women's reproductive health could not be assessed without them. Their speed, versatility, economy, specificity, and reliability are the critical qualities that make them useful in the evaluation of complex, dynamic events.

Analytic measures alone, however, are never sufficient to completely evaluate a complex system. There are many kinds of biomarkers, including many that are still to be recognized and properly used, and many kinds of indicators that we can use to monitor ecological health. The future of ecological health surveillance that I envisage is one where immunoassays are applied outside of the laboratory and data are immediately available on site. I will try to demonstrate that the future of field assessments lies within the realm of the immunoassays we have today. We can conceptualize procedures for tracking multiple events, perhaps not in real time, but in some reasonably short time interval, and for obtaining answers without instrumentation.

Let me return to and expand upon the original theme that I set forward in the beginning. I suggested using human reproductive health as a metaphor for ecological health, and I will now try to do this with five points.

1-56670-309-3/98/$0.00+$.50

First, reproductive health, at least for humans, is often unpredictable and somewhat variable. Reproduction is dynamic, and different individuals have very different expectations of it. Differences in expectations derive from divergent personal, social, or religious bases. Whatever the foundation for the basis of differences, expectations will vary between individuals, groups, and, often, for the same individual over time. In our ancestral state, human reproduction was most likely an ongoing process, limited only by the individual's inherent reproductive capacity within the context of the local environment. The advent of civilization brought the reduction in infant mortality and increased longevity which forced the recognition that this "natural" reproductive process could not remain unrestrained. The result is that human reproduction is now largely managed or regulated by drugs, traditions, or local economies. Of all of our natural traits, reproduction is probably managed more than any other. It's disappointing to us, therefore, when we discover that the methods we use for the convenience of managed reproduction may have side effects or health risks. Nonetheless, humans usually accept the risks as the cost for conformity, economy, and convenience.

While reproduction is critical for our long-term survival, it can be elective to any individuals at any one point in time and can be abandoned for prolonged time periods with little adverse effect. Like the expansion of the human population in the face of uncontrolled reproduction, ecological health cannot be left to the natural order of things when the pressure to "develop" all natural resources continues to increase. It is difficult to find an ecosystem that has not yet been, or does not continue to be, "managed" to some degree. Like reproductive health concerns, our concern about ecological health can be delayed through temporary or short-sighted management plans. While we recognize the potential side effects of many current management practices, the long-range risks are offset by short-range economic and political benefits.

Second, reproduction is complex and vulnerable to both internal as well as external change, physical and social order as well as natural and man-made events. All aspects of life impinge upon both the process and the outcome. From a functional point of view, reproduction involves the interaction of two physiologic systems and requires the interaction between at least two individuals. A change that affects one component may or may not have an effect on another component. Reproduction is downstream of all environmental inputs and is influenced by them. Since reproduction is not essential for short-term survival, it can be sacrificed for short time intervals as an adaptation to hostile environments. From a practical point of view, reproductive success is determined by our economy, religion, and interpersonal relationships. In one way or another all aspects of the changing environment impinge upon reproduction much as environmental changes influence ecological health.

Third, many of the events of primary importance in reproduction are concealed from us, and many of the activities associated with reproduction have other functions. Productivity is more apparent than the lack of it. We are not aware, for instance, when ovulation, fertilization, or implantation take place ... or when they fail. Recognizing reproductive failure is not a concern until reproductive success is desired. We tend to assess the complete reproductive process only in terms of the desired downstream product rather than in terms of a process on which the product depends.

This is often the case with ecosystems in that an immediate benefit takes precedence over the process that the system is dependent upon. A management plan may function to preserve the immediate, more visible, and aesthetic aspects of a portion of an ecosystem without proper attention to the long-range survival or natural productivity of the whole ecosystem. We often fail to recognize or understand the inter-relationships of the complete ecosystem, particularly the processes that are concealed or simply less obvious. We often forget to monitor processes that provide no immediate product but which reflect the general health of an operating system.

Fourth, reproductive health and reproductive potential are difficult to assess and impossible to predict. In terms of male reproductive potential or male fecundity, we have no single objective evaluation that can predict reproductive potential. We have no single indicator or set of parameters that tells us that a male will or will not be able to reproduce. The definition of male infertility, for example, is often based on the length of time and effort a couple must invest in trying to produce a child. A couple must try for at least one year before they are considered to be less than normal. This "progress along a time line" is not unlike the way we evaluate ecological health and the viability of ecosystems. Because of the delay in recognizing health problems in ecosystems, many ecosystems often deteriorate further during the process of evaluation. In the extreme, the "natural" health of an ecosystem is never recognized.

Fifth, and finally, the agencies concerned with human reproductive health have made an observation: infertility or the involuntary lack of reproduction has increased over the last four years. The experts in reproductive health say that perhaps as much as 37% of this infertility that we are seeing in the human population is caused by anthropogenic environmental factors. I think it is possible this is a corollary of what we see as "ailing" ecological health and may be more than just a metaphor.

If such a metaphor is useful, then it should provide insight into defining ecological health, understanding the processes critical to it, and formulating plans of action. We should plan our strategy as epidemiologists design their studies; and I am only suggesting now that we pay particular attention to the activities of reproductive epidemiologists, the trends in human reproductive performance, and the methods by which human reproduction continues to be managed to meet the needs of a complex world.

Part Three

Future Methods in Ecotoxicology

13 Future Directions for Ecotoxicological Methods

James N. Seiber

It is fitting to assess where we are in the development of ecotoxicology from the perspective of methodology which can support ecotoxicology research. Ecotoxicology is still an emerging field of study, moving very rapidly and attracting a lot of public attention, particularly in regard to what it might be able to contribute to risk assessment. It is, without question, a field of the future. We need ecotoxicological methods that are as sensitive as our analytical chemical methods. Right now there is a gap between what the chemists can detect and what the ecotoxicologists can interpret, but it doesn't have to be that way. Extremely sensitive biological responses exist, like moth antennae responding to pheromones and cholinesterase enzyme responding to activated organic phosphates. There is no question that such biological systems are extremely sensitive, so we should aim to reduce the gap between analytical chemistry and ecotoxicology. We also need methods that have more relevance to the real environment, not just the chamber or system to which the method owes its origin. This means that the methods have to be able to distinguish between a real response and the natural fluctuations in the environment, e.g., fluctuations due to weather and food supply. Finally, it is clear that we need methods that have a mechanistic basis so that we can find out how and why organisms sense the chemical exposure or exhibit some chemically-induced stress. To deliver methods that have these three characteristics — sensitivity, relevance, and a mechanistic basis — is going to require collaboration. For example, the analytical chemist must work closely with the field ecologist, and the biochemical toxicologist must work closely with the statistician.

The University of California, Davis, provides five centers: the Ecotoxicology Program, the U.S. EPA Center for Ecological Health Research, the Superfund Program, the Agricultural Health and Safety Center, and the NIEHS Center of Excellence on Agrochemicals in an unusual convergence of funds, faculty, and graduate students — just what it is going to take to foster the creativity needed for new developments. Interinstitutional collaborations (e.g., with the University of Nevada, Reno, the University of California, Santa Cruz, and several other institutions) open the way for whole new technologies. The institutional boundaries should be regarded as artificial and something to be minimized. Ron Tjeerdema writes about the NMR probes for phosphate energy storage depots that he, Don Crosby, and others are working with. This should catch a lot of attention and eventual use. Bruce Hammock writes about immunoassays coupled with liquid chromatography, and Gary Cherr describes developmental biomarkers in marine invertebrates. These techniques hold much promise.

1-56670-309-3/98/$0.00+$.50

What is needed is to take another step and to gaze into the future regarding extensions of these ideas and whole new approaches. In my own case, I would make a pitch for integrated methods — methods that have both an analytical and a bioassay component which can address the gap between the two approaches. Then, when we have a slice of the environment (e.g., fogwater contaminated with chemicals, rainwater, river water), we can describe its significance by integrating the biological responses with analytical chemistry. The hawk in the orchard is a good example of a biological integrator because of its exposure to chemicals and its demonstration of deleterious effects. It can also be worked with analytically so we can get this multimedia/exposure pathway integration, looking at what the hawk breathes in the air, what it takes in its food supply, what it contacts with its talons when it's roosting on the branches, and what deposits on its feathers. The analytical chemist can measure those exposures, and biochemical toxicologists can tell us what they mean.

A lot of headway can be made with a good model, but what happens when we get out to pristine environments? The hawk in the orchard is being sprayed, so we have an identifiable source of chemical stressors. What happens when we get out into the Sierra Nevada mountains or any other remote location and start to ask those questions? We can still measure the chemicals, although concentrations are typically very low. Can we be content that the levels are insignificant? A mountain landscape can integrate a lot of pollutants that might impinge on the neighboring watershed and collect in the drainage lake. We need tools that will allow us to assess the significance of those pollutants that are involved in the deposition–runoff process. We also need to be looking at other organisms that can be added to this evaluation scheme. No single organism is going to answer all the questions, but this is the kind of tool that we need to adapt to environmental situations.

We're interested, as I mentioned, in integrating bioassay and chemical assay. A supercritical fluid extractor solves a major problem for us. It helps avoid, or at least minimizes, the organic solvent that can cause complications. You extract a sample and then want to get the concentrated extract into a bioassay; there is all the organic solvent plus the contaminants that the solvent introduces getting in the way. The supercritical fluid extractor uses carbon dioxide as the extractant. It is also portable and can be used in the field. And wouldn't it be great if we could do a lot of sample preparation and screening in the field and not have to wait a few weeks or months for the answer to come back from the analytical lab? Of course, if you want to go far into the future, it would be even better if we had analytical methods that gave us real-time responses out in the field. That's the way organisms can respond; it would be nice to have those two pieces of data coming in simultaneously, in real time, continuously.

My bottom line is, let's look for methods that integrate analytical chemistry with ecotoxicology. Looking way down the road, it would be nice if we had an autoanalyzer-type system that would take the sample, extract it, and divert the extract a couple of ways so we get both biological data and chemical analysis on the same sample and maybe even biological data from a number of either whole organisms or enzymatic endpoints. This takes the Microtox® and Unitox®, which are steps in this direction, to a further progression.

14 Development and Application of Immunoassays for Biological and Environmental Monitoring

Adam S. Harris, Ingrid Wengatz, Monika Wortberg, Sabine B. Kreissig, Shirley J. Gee, and Bruce D. Hammock

INTRODUCTION

One ecosystem stressor of great scientific and public concern is the impact of toxic chemicals. Over the last 25 years, this concern has grown significantly because of the dramatic ecological and human health effects caused by a few of the more toxic chemicals that have been released into the environment. Of particular concern, obviously, are pesticides, compounds that are designed specifically to kill organisms.

Ecologists, wildlife biologists, toxicologists, and other environmental scientists devote much effort to the study of pesticides in the environment. Although most of these chemicals have specific, known biological effects, it is not easy to predict their effects on ecosystems once they are released. The majority of our knowledge of the effects of pesticides has been gathered from controlled laboratory studies where the molecular, cellular, and/or organismal level effects are often relatively easy to discern. The effects of the same pesticides on communities and whole ecosystems are extremely difficult, if not impossible, to study under such defined, controlled conditions. However, by monitoring the movement of pesticides through the environment and through organisms, populations, and communities, environmental scientists can study the ecosystem level impacts of pesticides and other toxic chemicals. The demonstration of cause and effect relationships between the presence of these chemicals and observed changes in ecosystems, through the use of environmental and biological monitoring, demands sound analytical chemistry.

Unfortunately, most classical analytical methods, such as gas chromatography (GC) and high pressure liquid chromatography (HPLC) are not well suited for the requirements of biological and environmental monitoring studies, which frequently involve a tremendously large number of samples. Classical methods are limited by low sample throughput, a characteristic of any method that requires serial analysis of samples, and high cost per sample (into the hundred dollar range for some analytes). These expensive technologies often put the field biologist in the position of having to contract for analytical assistance or to develop collaborations with laboratories where field data do not receive the highest priorities.

Immunoassay, an analytical method fairly new to the environmental sciences, offers an appropriate analytical option for many monitoring studies. It is a tool that environmental scientists and field biologists should utilize more than they currently do. Immunoassays can provide high-quality analytical data at a fraction of the cost of competing methods. The economies of scale with immunoassays lend themselves to large ecosystem projects. The low startup costs and simplicity of the assays allow them to be performed in biology laboratories. Also, immunoassays can be formatted to allow them to be performed at remote locations or even in the field. The speed and portability of immunoassays provide environmental scientists and field biologists with data to make instant decisions regarding further sampling or even remediation.

In this chapter, we will review the development and application of immunoassays for biological and environmental monitoring with emphasis on the research conducted in this laboratory.

ADVANTAGES AND LIMITATIONS OF IMMUNOASSAY FOR MONITORING

Over the past three decades, radioimmunoassay (RIA) and later enzyme immunoassay (EIA) have become important analytical techniques, particularly in biological research and clinical medicine. Although there have been many variations of immunoassay introduced since the pioneering work of Yalow and Berson in the late 1950s (Yalow and Berson 1959), the foundation of all variations is the same: antibodies have high affinity for their complementary antigens. It is this affinity that is exploited in all immunoassays. Any analyte to which a high-affinity antibody can be generated can most likely be detected by some form of immunoassay.

Advantages of immunoassay have been reviewed numerous times (Hammock and Mumma 1980; Oellerich 1980; Hammock et al. 1987; Cheung et al. 1988; Vanderlaan et al. 1988; and Harris et al. 1995). Briefly, advantages include:

1. Low detection limits
2. High analyte selectivity without chromatographic steps
3. Minimal sample preparation
4. High throughput of samples
5. High adaptability and easy coupling to other techniques
6. Applicability to many analytes
7. Cost effectiveness for monitoring programs

EIA has the additional advantage over RIA of not requiring the use of radioisotopes as labels and is, therefore, better suited for widespread use and field applications. By listing these advantages, we do not mean to imply that no other analytical method meets or exceeds the capability of immunoassay in some of these areas. Certainly some methods do. Taken together, however, the advantages make immunoassay a powerful analytical technique, especially for environmental and biological monitoring.

As is the case with every analytical method, immunoassay has limitations. It is as important to understand these limitations as to know the advantages. The limitations of immunoassay have been described previously in detail (Hammock et al. 1990). Briefly, key limitations are:

1. Longer assay development time than some classical analytical techniques
2. Interferences from sample matrices and from analyte analogs
3. Perceived as only for the determination of known single analytes
4. Low availability of critical reagents
5. Current lack of use in environmental analytical laboratories

As with the advantages listed above, we do not mean to imply that no other analytical method suffers to some degree from these limitations. Some methods do. In addition, these limitations rarely exclude immunoassay as a potential method for a given analytical problem. Developers and users of immunoassays are making progress in dealing with all of these limitations.

In general, immunoassays are easier to develop if the target analyte is fairly large, hydrophilic, stable, nonvolatile, and foreign to the host animal (Hammock et al. 1990). However, immunoassays have been developed to chemicals that do not fit all of these characteristics. Examples include immunoassays for relatively small molecules (less than 200 molecular weight) such as 4-nitrophenol (Li et al. 1991b), 1-naphthol (Krämer et al. 1994b), and monuron (Karu et al. 1994 and Schneider et al. 1994); hydrophobic chemicals such as dioxins (Stanker et al. 1987); hydrolytically unstable chemicals such as carbaryl (Marco et al. 1993a); and volatile chemicals such as thiocarbamates (Gee et al. 1988).

EIA is not the one analytical method for all analytes in all situations. For the analysis of most volatile organic compounds, GC will remain the method of choice. EIA is, however, an excellent tool in any analyst's repertoire because it complements the classical methods, thereby offering secondary methods for some compounds and the only reasonable choice for others. Most EIAs can be used to obtain quantitative results of equal or greater accuracy, precision, and sensitivity than other methods. However, one of the most appropriate initial uses for EIAs in an environmental laboratory is for screening large numbers of samples to identify positives which can then be analyzed by classical methods or by a more quantitative immunoassay procedure. Other uses in the environmental field for which immunoassay is well suited include monitoring for specific chemicals known to cause an observed effect in wildlife and monitoring concentrations of toxic chemicals following the addition or removal of a point source or during waste site clean-up.

STEPS IN IMMUNOASSAY DEVELOPMENT

In this laboratory, many immunoassays have been developed for compounds of interest to environmental chemists, ecologists, wildlife biologists, and toxicologists, especially for herbicides and insecticides. The procedures used to develop immunoassays have been reviewed extensively (Tijssen 1985; Harrison et al. 1988; Jung et al. 1989; Hall et al. 1990; and Nugent 1992). The steps below describe, in general terms, the process by which all enzyme immunoassays in this laboratory are developed.

Target Analyte Selection

Because the development of immunoassays requires a significant investment of time and resources, target analyte selection is important. Selection is based on analyte stability in the matrices of interest, the potential utility of the immunoassay once developed, and the need for either a compound-specific assay or one that detects the members of a class of closely related compounds.

Hapten Design and Synthesis

Haptens are target analyte mimics which are used both for the immunization of animals and for competitive binding in the assays. Ideal haptens have close chemical similarity to the target analyte and possess a functional group for coupling to proteins. Hapten retention of the analyte's unique functional groups, especially groups that hydrogen bond or are charged, is critical for the production of high-affinity antibodies that will lead to an immunoassay of high selectivity and sensitivity. The functional group used for coupling to proteins should be distal to the unique features of the target analyte, thus maximizing exposure of these unique features to the immune system of the host animal. Also important are the ease of hapten synthesis, hapten solubility, and the nature of the method to be used for conjugation to proteins. Three detailed reviews of the strategies of hapten design and synthesis used in this laboratory have been published (Goodrow et al. 1990; Harrison et al. 1991b; and Goodrow et al. 1995).

Hapten–Protein Conjugation

In order to elicit an immune response, small haptens must be conjugated to large carrier proteins. Highly immunogenic proteins with ample functional groups for coupling are the best choice for use as carrier proteins. In addition, the hapten used for immunization and other similar haptens are conjugated to enzymes and other proteins for use in the assays. For these hapten–protein conjugates, protein solubility, the presence of functional groups, and stability under reaction conditions (e.g., retention of enzyme activity) are important (Erlanger 1980). Many conjugation methods exist, and the selection of one is ultimately dependent on the hapten's available functional group. Conjugation methods have been reviewed elsewhere (Erlanger 1980; Kabakoff 1980; Tijssen 1985; and Brinkley 1992). Once the conjugation reaction is complete, unreacted compounds are separated from the hapten–protein conjugate by dialysis, gel filtration, or other methods (Kabakoff 1980; Tijssen 1985; and Brinkley 1992). Because hapten density is important for both immunization and assay performance, conjugation should be confirmed and hapten density determined by proven methods (Hillenkamp and Karas 1991; Brinkley 1992; and Wengatz et al. 1992).

Immunization

Because animal selection is partly dependent upon the desire for either polyclonal or monoclonal antibodies, this is decided prior to immunization. In this laboratory, we employ polyclonal antibody production in rabbits for most applications. Rabbits are immunized with the immunogen suspended in a mixture of a phosphate buffer and an adjuvant, which is a compound that enhances the immune response. Boost injections are given every three to four weeks, and small blood samples are drawn 10 days after each boost to monitor antibody production. Many other immunization protocols have been used (Tijssen 1985; and Harlow and Lane 1988). When the antibody titer reaches a useful level, the animals are sacrificed by exsanguination. The blood is centrifuged and the serum removed and stored with sodium azide at 4°C. Some researchers prefer to remove the antibodies from the whole serum and then work only with this purified antibody fraction (Tijssen 1985; and Harlow and Lane 1988).

Serum Characterization

The relative affinity of the antibodies for the immunizing hapten and other haptens is evaluated to determine antibody titers against each and to identify potentially useful antibody–hapten combinations for further immunoassay development (Gee et al. 1988). Inhibition experiments are then conducted to determine the relative ease by which the target analyte can inhibit antibody–hapten binding. Inhibition experiments are also conducted with chemicals closely related to the target analyte to determine antibody selectivity.

Assay Development, Optimization and Validation

EIAs may be designed in several different formats, each with advantages and disadvantages. Two common formats are shown in Figure 14.1. Once a format is chosen, assay conditions, including incubation times, temperature, pH, and buffer and salt concentration, are optimized so that the assay performs as desired. The goal of optimization is an assay with the desired selectivity, sensitivity, and lower limit of detection. Matrix effects are then carefully examined. The effects of pH, salts and metal ions, organic solvents, and the components of biological and environmental samples are characterized. Sample preparation schemes are devised if the assay is not rugged enough in matrices of interest. Samples often can be diluted in an aqueous buffer to eliminate matrix effects. Finally, the immunoassays are validated with spike-recovery studies using water, soil, urine, blood, and/or other environmental and biological samples, just as all new analytical methods are validated. Occasionally during the development and optimization of a new EIA or during the application of an existing EIA, unusual and/or irreproducible results are obtained because of complex sample matrices, the degradation of reagents, or other problems. A recent publication describes in detail rational troubleshooting steps for both novice and experienced EIA users (Schneider et al. 1995).

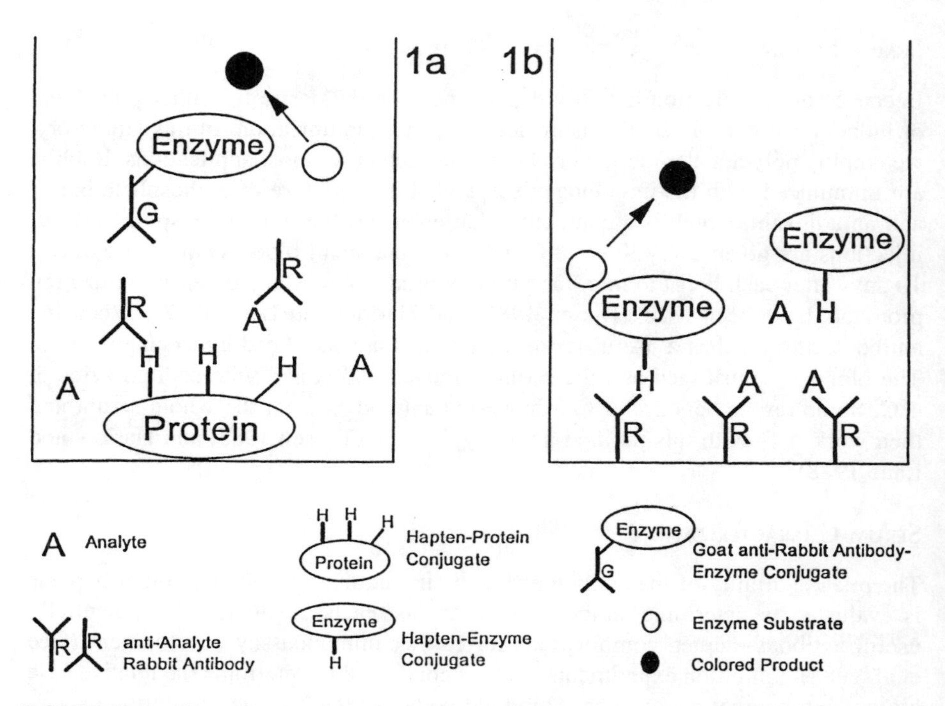

FIGURE 14.1 General scheme of two enzyme immunoassay formats. Both formats can be run in 96-well microtiter plates and in plastic tubes. In the coating antigen format (1a), anti-analyte rabbit antibodies bind to either the analyte, free in solution, or to the hapten of the hapten-protein conjugate, previously bound to the solid phase. Following a washing step which removes all material not bound to the solid phase, goat anti-rabbit antibody-enzyme conjugate is added. This conjugate will bind to anti-analyte rabbit antibody bound to the hapten-protein conjugate. Following another washing step, enzyme substrate is added, and then the colored product is detected. More analyte initially present results in less colored product. In the enzyme-tracer format (1b), anti-analyte rabbit antibodies, previously bound to the solid phase, bind either the analyte or the hapten of the hapten-enzyme conjugate. Following a washing step, enzyme substrate is added, and then the colored product is detected. As in the coating antigen format, more analyte initially present results in less colored product.

EXAMPLES OF IMMUNOASSAYS

As mentioned above, enzyme immunoassays have been developed in this laboratory for the biological and environmental monitoring of many compounds of environmental relevance: herbicides, insecticides, and other toxic compounds and some selected metabolites. The following examples are reviewed briefly to illustrate the range of small molecules for which immunoassays can be developed and some of the characteristics of these assays.

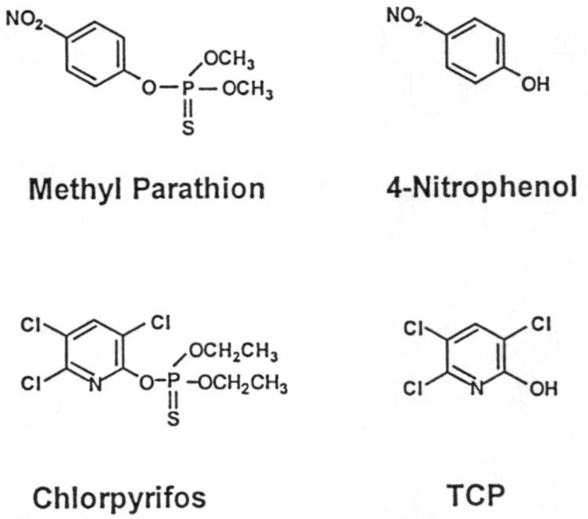

Methyl Parathion

4-Nitrophenol

Chlorpyrifos

TCP

FIGURE 14.2 The structures of target analytes for our OP project.

INSECTICIDES

Organophosphorous Insecticides and Metabolites

The organophosphorous (OP) insecticides were the first major class to replace the more persistent organochlorine class and still are widely used. The target analytes for this project are selected metabolites of some common OP insecticides: The 4-nitrophenol metabolites of five different OPs (methyl and ethyl parathion, EPN, dicapthon, and fenitrothion) and the 3,5,6-trichloro-2-pyridinol (TCP) metabolite of chlorpyrifos (Figure 14.2). The limit of detection for the 4-nitrophenol immunoassay is as low as 0.5 ng/mL, depending on the matrix. It has been applied for the analysis of 4-nitrophenols and parathion (following a hydrolysis step) in natural waters, soils, urine, and foods (Li et al. 1991b; and Wong et al. 1991). The TCP immunoassay is currently under development. By directing these assays toward the products of hydrolysis of the parent OPs, the assays can be used for detecting both the parent OPs and the metabolites in biological and environmental monitoring studies.

Pyrethroids

The pyrethroids are synthetic analogs of the natural pyrethrin insecticides, derived from flowers of the genus *Chrysanthemum*. The fact that some pyrethroids are as much as 5000 times more toxic to insects than to mammals shows their high species selectivity. Unfortunately, some pyrethroids are also toxic to some nontarget organisms, including fish. Because of their high insecticidal potency, they are used in very low concentrations in agriculture, in forestry, and for household purposes. Our primary targets are fenvalerate, fenpropathrin, and selected metabolites (Figure 14.3). A large series of potentially useful haptens has been synthesized. Currently, polyclonal sera are being raised and evaluated for assay development,

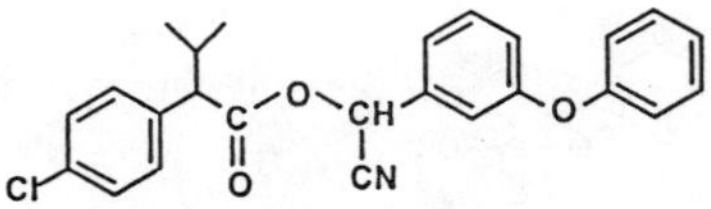

Fenvalerate

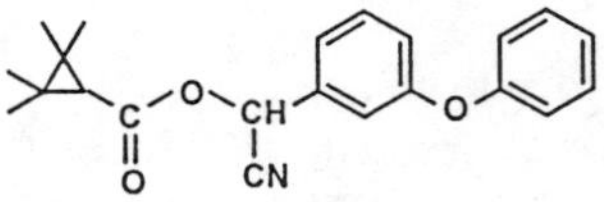

Fenpropathrin

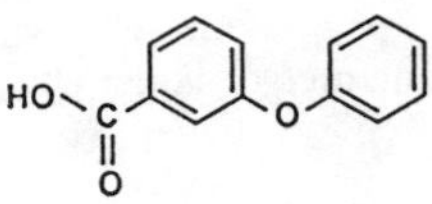

3-Phenoxybenzoic Acid

FIGURE 14.3 Structures of two pyrethroids and 3-phenoxybenzoic acid, a common metabolite.

which is a challenge because of the hydrophobicity of fenvalerate and fenpropathrin. Assays to both parent compounds and metabolites, as in the case of the OPs, will provide even better tools for monitoring pyrethroids in the environment.

Carbaryl

Carbaryl is one of the most common carbamate insecticides used in agriculture. It is applied to many fruit, vegetable, and field crops, but concern has arisen because of its adverse effects on bee populations. An assay based on polyclonal antibodies has been fully developed and characterized despite the challenge of carbaryl's hydrolytic instability. It has been applied for the analysis of soil, water, foods, and urine with a detection limit as low as 50 pg/mL in some sample matrices (Marco et al. 1993a). Another assay was developed for the major carbaryl breakdown product, 1-naphthol (Figure 14.4). This polyclonal antibody-based assay has a detection limit of 10 ng/mL (Krämer et al. 1994b).

Herbicides

s-Triazine Herbicides and Metabolites

The *s*-triazine class of herbicides is the most heavily used worldwide and presents environmental contamination problems in many areas. These compounds are excellent markers of agrochemical pollution. In general, if *s*-triazines are found in an environmental sample, other agrochemicals may be found also. If *s*-triazines are not found, then it is unlikely that other agrochemicals will be found. The target analytes

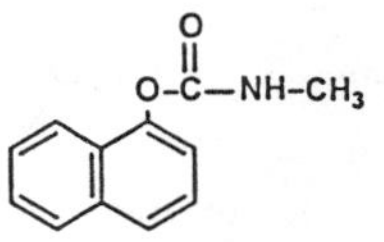

O-C(=O)-NH-CH$_3$

Carbaryl

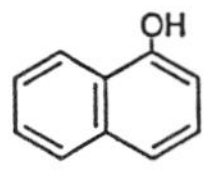

OH

1-Naphthol

FIGURE 14.4 Structures of carbaryl and 1-naphthol.

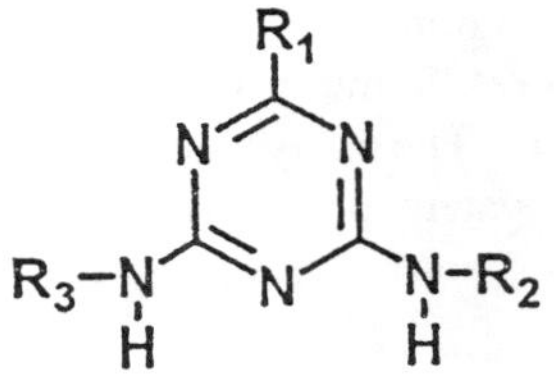

R_1, N, N, R_3–N(H), N, N(H)–R_2

Parent Triazines	R_1	R_2	R_3	Reference
Atrazine	Cl	ethyl	isopropyl	Harrison et al. 1991a; Lucas et al. 1991; Schneider & Hammock 1992
Simazine	Cl	ethyl	ethyl	Harrison et al. 1991a; Lucas et al. 1991
Ametryne	S-methyl	ethyl	isopropyl	Harrison et al. 1991a
Metabolites				
Hydroxyatrazine	OH	ethyl	isopropyl	Lucas et al. 1993b
Hydroxysimazine	OH	ethyl	ethyl	Lucas et al. 1993b
Deisopropylatrazine	Cl	ethyl	H	in progress
Deethylatrazine	Cl	H	isopropyl	in progress
Atrazine mercapturic acid	*N*-acetyl-L-cysteine	ethyl	isopropyl	Lucas et al. 1993a

FIGURE 14.5 Structures of selected analytes for the *s*-triazine project.

for this long-term project include several of the most common *s*-triazines and their metabolites (Figure 14.5). From a series of *s*-triazines haptens, some high-quality polyclonal and monoclonal antibodies were produced, and useful assays for several parent *s*-triazines and some metabolites have been developed (Harrison et al. 1991a; Lucas et al. 1991; Schneider and Hammock 1992; Lucas et al. 1993a; and Lucas

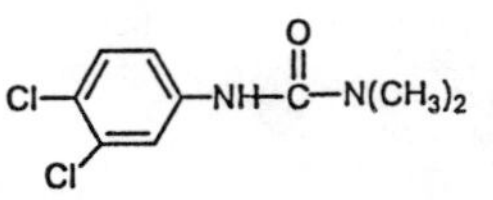

Diuron

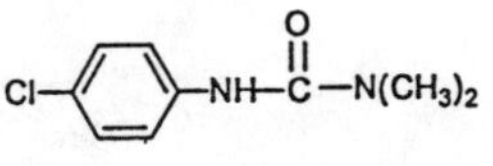

Monuron

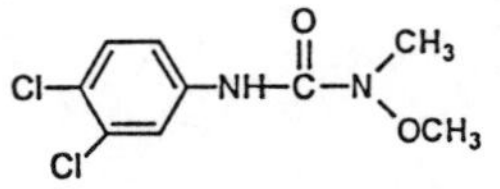

Linuron

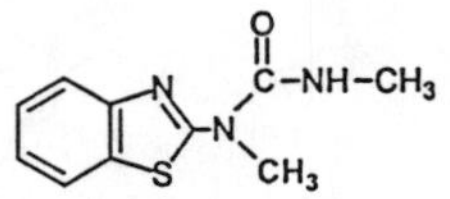

MBT

FIGURE 14.6 Structures of selected urea herbicides.

et al. 1993b). The limit of detection for one of these assays is as low as 30 pg/mL (Schneider and Hammock 1992). Some of the assays have high tolerance for urine and organic solvents as matrices. The assays have been applied to determine triazines and/or metabolites in natural waters, soils, urine, and foods.

Urea Herbicides

Immunoassays for some of the arylurea herbicides (diuron, monuron, and linuron) have been developed (Figure 14.6). These assays can be used for both class-specific and compound-specific analysis in a variety of samples. The detection limits for these assays are as low as 40 pg/mL with polyclonal antibodies (Schneider et al. 1994) and 600 pg/mL with monoclonal antibodies (Karu et al. 1994). Another urea herbicide, methabenzthiazuron (MBT), provided an excellent target for immunoassay development because of its distinctive structure (Figure 14.6). A polyclonal antibody-based assay in the enzyme-tracer format had a detection limit of 30 pg/mL in water samples (Kreissig et al. 1991; and Kreissig and Hock 1993).

Triclopyr

A recent development is an immunoassay for the herbicide triclopyr (Figure 14.7), used for post-emergence control of many broad-leaf and woody plants. Heavy use of triclopyr in forestry for site preparation and conifer release has justified improved methods of triclopyr determination in environmental samples. The lower detection limit of this polyclonal antibody-based assay is below 1 ng/mL, and spike-recovery studies in natural water samples suggest that this assay will be a valuable tool for rapid, inexpensive monitoring.

In addition to those described above, enzyme immunoassays for other herbicides have been developed in this laboratory: paraquat (Van Emon et al. 1986), thiobencarb, molinate (Gee et al. 1988), glyphosate, bentazon (Li et al. 1991a), and bromacil (Szurdoki et al. 1992; and Bekheit et al. 1993).

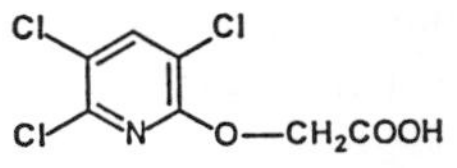

Triclopyr

FIGURE 14.7 Structure of triclopyr.

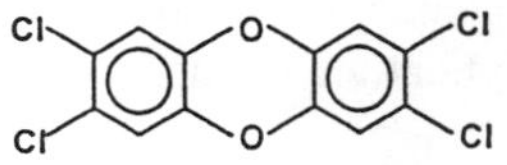

2,3,7,8-Tetrachlorodibenzodioxin (TCDD)

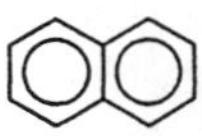

Naphthalene

FIGURE 14.8 Structures of TCDD and naphthalene.

Other Toxic Chemicals, Metabolites, and Adducts

In addition to immunoassays for pesticides, this laboratory has begun development of assays for other toxic compounds that pose a threat to ecosystems, wildlife, and humans. Although the area is less advanced than the pesticide area, several immunoassays for industrial chemicals and pollutants have been developed by others. Two examples are immunoassays for polychlorinated dibenzodioxins and dibenzofurans (Stanker et al. 1987) and for trinitrotoluene and other nitroaromatics (Eck et al. 1990). In this laboratory, current target analytes include 2,3,7,8-tetrachlorodibenzodioxin (TCDD) and metabolites and adducts of naphthalene (Figure 14.8). TCDD is well known as a toxic contaminant, produced as a side product in some chemical syntheses and processes and during waste incineration. TCDD is extremely potent, especially as a carcinogen in some species, and therefore a rapid, inexpensive immunoassay with a very low detection limit would be particularly desirable for its detection. The same antibodies could be used for affinity concentration and purification systems before classical or immunochemical analysis. Naphthalene is used in the synthesis of many chemicals, including dyes, resins, and oil additives, and appears as a toxic contaminant of the environment from several sources, such as automobile exhaust, coal combustion, and coal gasification.

Because of the physical and chemical properties of these analytes, in particular their hydrophobicity, EIA development is difficult and requires novel approaches to hapten synthesis. In the case of naphthalene, EIAs for naphthalene, major metabolites, and adducts have been or are being developed, thus providing a set of analytical tools for the study of this compound in the environment and its biological effects (Marco et al. 1993b; and Marco et al. 1993c).

NEW APPLICATIONS FOR IMMUNOASSAYS AND ANTIBODIES

Compared to most other analytical methods used in environmental sciences, immunoassay is relatively new and evolving rapidly. There are tremendous possibilities for future uses of both immunoassays and antibodies in other formats that may have a great influence on the analytical techniques used by environmental scientists.

NEW TARGET ANALYTES

There are many new targets for immunoassay development that would be of benefit to any scientist interested in monitoring ecosystems for the impact of synthetic chemicals and other stressors. These potential targets include the majority of pesticides and other industrial chemicals, for which immunoassays do not exist, and molecules that can serve as biomarkers of exposure or effect, such as phase II metabolites, metallothionein, and stress proteins. New targets for exposure monitoring by immunoassay have been discussed recently (Harris et al. 1995). Another exciting area is the development of rapid, field portable assays for toxic metal ions. For example, immunoassays, as illustrated by immunoassays for mercuric ions (Wylie et al. 1991; 1992) and other ligand-based assays analogous to immunoassays (Szurdoki et al. 1995) for metal ions are in progress.

MULTIANALYTE METHODS

Immunoassay traditionally has been used as a single-analyte method, and this often is noted as a great limitation of the method. With carefully designed haptens it is possible to generate both monoclonal and polyclonal antibodies which are capable of detecting a class or subclass of compounds. Class-selective assays are now in use in some laboratories and on the market for a small number of environmentally relevant compounds. Quantitative immunoassay-based multianalyte methods are a newer area of research. This work usually involves combining antibodies produced for different analytes in one assay or in parallel assays on one microtiter plate and a mathematical evaluation of the data. Current progress suggests that such methods will be successfully developed and validated (Muldoon et al. 1993; and Jones et al. 1994). These advances will offer more opportunities for environmental scientists to use immunoassay for the analysis of samples containing unknown chemical pollutants.

RECOMBINANT ANTIBODIES

Several laboratories are using the techniques of molecular biology to produce, and hopefully improve, the antibodies or antibody fragments used for immunoassay and for other purposes. The advantages that these recombinant antibodies offer are numerous:

1. The ability to manipulate and optimize binding characteristics for altered affinity and tailored specificity
2. The ability to engineer into the antibody sequence additional features such as reporter enzymes, metal binding sites, and affinity tags
3. Rapid production of antibodies requiring the sacrifice of fewer animals
4. Low-cost, large-scale production of antibodies

Currently, researchers are investigating different prokaryotic and eukaryotic expression systems, a variety of sizes of the expressed protein, and the necessary conditions for high expression levels. Recombinant antibody research in this laboratory has been described in recent publications (Ward et al. 1993; Choudary et al. 1995; Ward et al. 1995a,b; and Kreissig et al. 1995).

Immunoaffinity Chromatography and Hyphenated Techniques

Some very promising uses for antibodies in the routine analysis of biological and environmental samples involve immunoaffinity chromatography and hyphenated analytical techniques. In these techniques, the antibodies are used either for sample clean-up, as in immunoaffinity chromatography, or for analyte detection, as in coupled techniques like HPLC-EIA. Immunoaffinity chromatography requires fixing antibodies onto column packing material and then passing samples through under low pressure. An elution step follows to release the antibody-bound analyte. Using this technique, analytes and interfering compounds can be separated efficiently from very complex sample matrices, such as blood. In addition, analytes often can be concentrated prior to detection by another method, and in some cases, immunoaffinity chromatography can be used alone as an analytical method (Kim et al. 1993; and Wortberg et al. 1994a). The coupling of immunoassay or immunoaffinity chromatography with another technique offers the analyst the advantages of both methods. Examples include solid-phase extraction-EIA (Aga and Thurman 1993), supercritical fluid extraction-EIA (Wong et al. 1991), HPLC-EIA (de Frutos and Regnier 1993; and Krämer et al. 1994a), immunoaffinity chromatography-GC (Farjam et al. 1991), and immunoaffinity chromatography-LC-MS (Rule et al. 1994). In-line coupled techniques offer the additional advantage of rapid, continuous analysis of complex environmental and biological samples.

Dipstick Devices

One format for rapid field analysis is the disposable dipstick format. This format has been applied to the detection of clinical targets (e.g., hormones and blood sugar) and food contaminants, but is relatively new to the environmental field. There are variations of this concept, but all involve procedures with a few, very short steps and visual determination. Although this format produces qualitative rather than quantitative results, the potential advantages make this an attractive format for specific applications, especially as a screening tool for water and urine samples. Recent publications illustrate two variations of the dipstick concept for the detection of herbicides (Siebert et al. 1993; and Giersch 1993).

Biosensors

A large number of biosensors, which combine the specificity and sensitivity of immunoassays with physical signal transduction, have been developed in recent years for environmental and clinical analytes. The three components of a classical biosensor are a receptor (e.g., an antibody, binding protein, or enzyme), a transducer (e.g., an optical fiber or electrode), and signal processing electronics. The receptor is usually attached (immobilized) to the transducer surface, which enables it to detect interaction with analyte molecules. In contrast to immunoassays, biosensors rely on the reuse of the same receptor surface for several measurements. Direct signal generation potentially enables real-time monitoring of analytes, thus making biosensors suitable tools for continuous environmental monitoring. Parallel to the increase in the number of immunoassays to detect compounds of environmental relevance, some biosensor research also has made this shift (Vo-Dinh et al. 1987; Anis et al. 1992; Bier et al. 1992; and Wortberg et al. 1994b). However, in contrast to immunoassay test kits, few biosensors are commercially available. As immunoassay becomes more established, the acceptance of biosensors as future alternatives to classical analytical techniques will increase.

Antibody-Based Clean-up Systems

Because of the high affinity of antibodies for their complementary antigens, antibodies someday might become the basis of selective clean-up systems, especially for the removal of chemical contaminants from water. Assuming a great supply of inexpensive antibodies or stable antibody fragments and efficient elution procedures, antibodies selective for a toxic chemical could be immobilized on polymer beads, placed in contaminated water, and later removed, thus reducing the concentration of the chemical in the water. Large-scale column clean-up systems might also be possible. Although enzymes have been suggested as being better for such applications, antibodies may have value as well. In addition, antibodies with a high affinity for an organic pollutant could be engineered into a microbe or plant so that the organism acts to bioaccumulate the pollutant. Progress also is being made to use recombinant technology to engineer catalytic activity into antibody molecules (Lerner et al. 1991; and Lesley et al. 1993).

CONCLUSIONS

The development of enzyme immunoassays for pesticides and other compounds in this laboratory has led to many rapid, inexpensive, and sensitive analytical methods which are well suited for environmental and biological monitoring. The fully developed assays have been validated with a variety of environmental and biological samples, and some have already been used in monitoring studies.

This work has led to a greater understanding of the properties of synthetic haptens which produce high-affinity antibodies and of the better immunoassay formats for rapid, rugged, and inexpensive assays. Such knowledge allows future assays to be developed more efficiently. Success with some of the current challenging projects (e.g., TCDD) will broaden the range of compounds for which immunoassay may be an appropriate analytical method. Additionally, assays for the determination of classes of compounds, instead of single analytes, and methods of multianalyte quantitation by immunoassay will provide better analytical tools for scientists in the environmental field. Many other promising new applications of antibodies in analytical chemistry and the environmental sciences also exist.

It is certain that immunoassays will be used more heavily in the future for routine environmental and biological monitoring. Therefore, it is important that scientists in many fields become familiar with immunoassay techniques and include them in their repertoire of analytical methods.

ACKNOWLEDGMENT

This research was funded by NIEHS Superfund 2P42-ES04699, U.S. EPA CR 819047, USDA Forest Service NAPIAP R8-27, Center for Ecological Health Research CR 819658, NIEHS Center for Environmental Health Sciences 1P30-ES05707, and the UC Davis Ecotoxicology Program. A. S. Harris was supported by a training grant from the UC Davis Ecotoxicology Program; M. Wortberg was supported by Deutsche Forschungsgemeinschaft and Fonds der Chemischen Industrie; and S. B. Kreissig was supported by the Gesellschaft fur Biotechnologische Forschung mbH and Bayer AG, Germany. B. D. Hammock was a Burroughs Wellcome Toxicology Scholar.

REFERENCES

Aga, D. S. and Thurman, E. M., Coupling solid-phase extraction and enzyme-linked immunosorbent assay for ultratrace determination of herbicides in pristine water, *Anal. Chem.*, 65, 2894, 1993.

Anis, A. N., Wright, J., Rogers, K. R., Thompson, R. G., Valdes, J. J., and Eldefrawi, M. E., A fiber-optic immunosensor for detecting parathion, *Anal. Lett.*, 25, 627, 1992.

Bekheit, H. K. M., Lucas, A. D., Szurdoki, F., Gee, S. J., and Hammock, B. D., An enzyme immunoassay for the environmental monitoring of the herbicide bromacil, *J. Agric. Food Chem.*, 41, 2220, 1993.

Bier, F. F., Stöcklein, W., Böcher, M., Bilitewski, U., and Schmid, R. D., Use of a fiber optic immunosensor for the detection of pesticides, *Sensors Actuators B*, 7, 509, 1992.

Brinkley, M., A brief survey of methods for preparing protein conjugates with dyes, haptens, and cross-linking reagents, *Bioconj. Chem.*, 3, 2, 1992.

Cheung, P. Y. K., Gee, S. J., and Hammock, B. D., Pesticide immunoassay as a biotechnology, in *The Impact of Chemistry on Biotechnology*, ACS Symposium Series 362, Phillips, M., Shoemaker, S. P., Middlekauff, R. D., and Ottenbrite, R. M., Eds., American Chemical Society, Washington, D.C., 1988, chap. 18.

Choudary, P. V., Lee, H. A., Hammock, B. D., and Morgan, M. R. A., Recombinant antibodies: new tools for immunoassays, in *New Frontiers in Agrochemical Immunoassay*, Kurtz, D. A., Skerritt, J. H., and Stanker, L., Eds., AOAC, Arlington, VA, 1995, 171.

de Frutos, M. and Regnier, F. E., Tandem chromatographic-immunological analyses, *Anal. Chem.*, 65, 17A, 1993.

Eck, D. L., Kurth, M. J., and Macmillan, C., Trinitrotoluene and other nitroaromatic compounds: immunoassay methods, in *Immunochemical Methods for Environmental Analysis*, ACS Symposium Series 442, Van Emon, J. M. and Mumma, R. O., Eds., American Chemical Society, Washington, D.C., 1990, chap. 9.

Erlanger, B. F., The preparation of antigenic hapten-carrier conjugates: a survey, *Methods Enzymol.*, 70, 85, 1980.

Farjam, A., Vreuls, J. J., Cuppen, W. J. G. M., Brinkman, U. A. T., and de Jong, G. J., Direct introduction of large-volume urine samples into an on-line immunoaffinity sample pretreatment-capillary gas chromatography system, *Anal. Chem.*, 63, 2481, 1991.

Gee, S. J., Miyamoto, T., Goodrow, M. H., Buster, D., and Hammock, B. D., Development of an enzyme-linked immunosorbent assay for the analysis of the thiocarbamate herbicide molinate, *J. Agric. Food Chem.*, 36, 863, 1988.

Giersch, T., A new monoclonal antibody for the sensitive detection of atrazine with immunoassay in microtiter plate and dipstick format, *J. Agric. Food Chem.*, 41, 1006, 1993.

Goodrow, M. H., Harrison, R. O., and Hammock, B. D., Hapten synthesis, antibody development, and competitive inhibition enzyme immunoassay for s-triazine herbicides, *J. Agric. Food Chem.*, 38, 990, 1990.

Goodrow, M. H., Sanborn, J. R., Stoutamire, D. W., Gee, S. J., and Hammock, B. D., Strategies for immunoassay hapten design, in *Immunoanalysis of Agrochemicals: Emerging Technologies*, ACS Symposium Series 586, Nelson, J. O., Karu, A. E., and Wong, R. B., Eds., American Chemical Society, Washington, D.C., 1995, 119.

Hall, J. C., Deschamps, R. J. A., and McDermott, M. R., Immunoassays to detect and quantitate herbicides in the environment, *Weed Technol.*, 4, 226, 1990.

Hammock, B. D. and Mumma, R. O., Potential of immunochemical technology for pesticide analysis, in *Recent Advances in Pesticide Analytical Methodology*, ACS Symposium Series 136, Harvey, J. and Zweig, G., Eds., American Chemical Society, Washington, D.C., 1980, chap. 18.

Hammock, B. D., Gee, S. J., Cheung, P. Y. K., Miyamoto, T., Goodrow, M. H., Van Emon, J., and Seiber, J. N., Utility of immunoassay in pesticide trace analysis, in *Pesticide Science and Biotechnology*, Greenhalgh, R. and Roberts, T. R., Eds., Blackwell Scientific, Ottawa, 1987, 309.

Hammock, B. D., Gee, S. J., Harrison, R. O., Jung, F., Goodrow, M. H., Li, Q. X., Lucas, A. D., Szekacs, A., and Sundaram, K. M. S., Immunochemical technology in environmental analysis: addressing critical problems, in *Immunochemical Methods for Environmental Analysis*, ACS Symposium Series 442, Van Emon, J. M. and Mumma, R. O., Eds., American Chemical Society, Washington, D.C., 1990, chap. 11.

Harlow, E. and Lane, D., *Antibodies: A Laboratory Manual*, Cold Springs Harbor Laboratory, Cold Springs Harbor, NY, 1988.

Harris, A. S., Lucas, A. D., Krämer, P. M., Marco, M. P., Gee, S. J., and Hammock, B. D., Use of immunoassay for the detection of urinary biomarkers of exposure, in *New Frontiers in Agrochemical Immunoassay*, Kurtz, D. A., Skerritt, J. H., and Stanker, L., Eds., AOAC, Arlington, VA, 1995, 217.

Harrison, R. O., Gee, S. J., and Hammock, B. D., Immunochemical methods of pesticide residue analysis, in *Biotechnology for Crop Protection*, ACS Symposium Series 379, Hedin, P. A., Menn, J. J., and Hollingworth, R. M., Eds., American Chemical Society, Washington, D.C., 1988, chap. 24.

Harrison, R. O., Goodrow, M. H., and Hammock, B. D., Competitive inhibition ELISA for the *s*-triazine herbicides: assay optimization and antibody characterization, *J. Agric. Food Chem.*, 39, 122, 1991a.

Harrison, R. O., Goodrow, M. H., Gee, S. J., and Hammock, B. D., Hapten synthesis for pesticide immunoassay development, in *Immunoassays for Trace Chemical Analysis: Monitoring Toxic Chemicals in Humans, Food, and the Environment*, ACS Symposium Series 451, Vanderlaan, M., Stanker, L. H., Watkins, B. E., and Roberts, D. W., Eds., American Chemical Society, Washington, D.C., 1991b, chap. 2.

Hillenkamp, F. and Karas, M., Matrix-assisted laser desorption-ionization mass spectrometry of biopolymers, *Anal. Chem.*, 63, 1193A, 1991.

Jones, G., Wortberg, M., Kreissig, S. B., Bunch, D., Gee, S. J., Hammock, B. D., and Rocke, D. M., Extension of the four-parameter logistic model for ELISA to multianalyte analysis, *J. Immunol. Methods*, 177(1-2), 1, 1994.

Jung, F., Gee, S. J., Harrison, R. O., Goodrow, M. H., Karu, A. E., Braun, A. L., Li, Q. X., and Hammock, B. D., Use of immunochemical techniques for the analysis of pesticides, *Pestic. Sci.*, 26, 303, 1989.

Kabakoff, D. S., Chemical aspects of enzyme-immunoassay, in *Enzyme-Immunoassay*, Maggio, E. T., Ed., CRC Press, Boca Raton, FL, 1980, chap. 4.

Karu, A. E., Goodrow, M. H., Schmidt, D. J., Hammock, B. D., and Bigelow, M. W., Synthesis of haptens and derivation of monoclonal antibodies for immunoassay of the phenylurea herbicide diuron, *J. Agric. Food Chem.*, 42, 301, 1994.

Kim, B. B., Vlasov, E. V., Miethe, P., and Egorov, A. M., Immunoaffinity chromatographic method for the detection of pesticides, *Anal. Chim. Acta*, 280, 191, 1993.

Krämer, P. M., Li, Q. X., and Hammock, B. D., Integration of liquid chromatography with immunoassay — an approach combining the strengths of both methods, *J. AOAC Int.*, 77(5), 1275, 1994a.

Krämer, P. M., Marco, M. P., and Hammock, B. D., Development of a selective enzyme-linked immunosorbent assay for 1-naphthol — the major metabolite of carbaryl (1-naphthyl N-methylcarbamate), *J. Agric. Food Chem.*, 42(4), 934, 1994b.

Kreissig, S., Hock, B., and Stöcker, R., An enzyme immunoassay for the determination of methabenzthiazuron, *Anal. Lett.*, 24, 1729, 1991.

Kreissig, S. and Hock, B., Entwicklung und Anwendung eines Enzymimmunoassays zur Bestimmung von methabenzthiazuron, in *Biochemische Methoden zur Schadstofferfassung im Wasser*, Fachgruppe Wasserchemie der GDCh., Weinheim, Germany, 1993, chap. 3.8.

Kreissig, S. B., Ward, V. K., Hammock, B. D., and Choudary, P. V., Sequence analysis of individual chains of antibodies reactive to triazine herbicides, in *Immunoanalysis of Agrochemicals: Emerging Technologies*, ACS Symposium Series 586, Nelson, J. O., Karu, A. E., and Wong, R. B., Eds., American Chemical Society, Washington, D.C., 1995, 31.

Lerner, R. A., Benkovic, S. J., and Schultz, P. G., At the crossroads of chemistry and immunology: catalytic antibodies, *Science*, 252, 659, 1991.

Lesley, S. A., Patten, P. A., and Schultz, P. G., A genetic approach to the generation of antibodies with enhanced catalytic activities, *Proc. Natl. Acad. Sci. U.S.A.*, 90, 1160, 1993.

Li, Q. X., Hammock, B. D., and Seiber, J. N., Development of an enzyme-linked immunosorbent assay for the herbicide bentazon, *J. Agric. Food Chem.*, 39, 1537, 1991a.

Li, Q. X., Zhao, M. S., Gee, S. J., Kurth, M. J., Seiber, J. N., and Hammock, B. D., Development of enzyme-linked immunosorbent assays for 4-nitrophenol and substituted 4-nitrophenols, *J. Agric. Food Chem.*, 39, 1685, 1991b.

Lucas, A. D., Schneider, P., Harrison, R. O., Seiber, J. N., Hammock, B. D., Bigger, J. W., and Rolston, D. E., Determination of atrazine and simazine in water and soil using polyclonal and monoclonal antibodies in enzyme-linked immunosorbent assays, *Food Agric. Immunol.*, 3, 155, 1991.

Lucas, A. D., Jones, A. D., Goodrow, M. H., Saiz, S. G., Blewett, C., Seiber, J. N., and Hammock, B. D., Determination of atrazine metabolites in human urine: development of a biomarker of exposure, *Chem. Res. Toxicol.,* 6, 107, 1993a.

Lucas, A. D., Bekheit, H. K. M., Goodrow, M. H., Jones, A. D., Kullman, S., Matsumura, F., Woodrow, J. E., Seiber, J. N., and Hammock, B. D., Development of antibodies against hydroxyatrazine and hydroxysimazine: application to environmental samples, *J. Agric. Food Chem.*, 41, 1523, 1993b.

Marco, M. P., Gee, S. J., Cheng, H. M., Liang, Z. Y., and Hammock, B. D., Development of an enzyme-linked immunosorbent assay for carbaryl, *J. Agric. Food Chem.,* 41, 423, 1993a.

Marco, M. P., Nasiri, M., Kurth, M. J., and Hammock, B. D., Enzyme-linked immunosorbent assay for the specific detection of the mercapturic acid metabolites of naphthalene, *Chem. Res. Toxicol.*, 6, 284, 1993b.

Marco, M. P., Hammock, B. D., and Kurth, M. J., Hapten design and development of an ELISA (enzyme-linked immunosorbent assay) for the detection of the mercapturic acid conjugates of naphthalene, *J. Org. Chem.*, 58, 7548, 1993c.

Muldoon, M. T., Fries, G. F., and Nelson, J. O., Evaluation of ELISA for the multianalyte analysis of s-triazines in pesticide waste and rinsate, *J. Agric. Food Chem.,* 41, 322, 1993.

Nugent, P. A., Enzyme-linked competitive immunoassay, in *Emerging Strategies for Pesticide Analysis,* Cairns, T. and Sherma, J., Eds., CRC Press, Boca Raton, FL, 1992, chap. 11.

Oellerich, M., Enzyme immunoassays in clinical chemistry: present status and trends, *J. Clin. Chem. Clin. Biochem.*, 18, 197, 1980.

Rule, G. S., Mordehai, A. V., and Henion, J., Determination of carbofuran by on-line immunoaffinity chromatography with coupled-column liquid chromatography mass spectrometry, *Anal. Chem.*, 66, 230, 1994.

Schneider, P. and Hammock, B. D., Influence of the ELISA format and the hapten-enzyme conjugate on the sensitivity of an immunoassay for *s*-triazine herbicides using monoclonal antibodies, *J. Agric. Food Chem.*, 40, 525, 1992.

Schneider, P., Goodrow, M. H., Gee, S. J., and Hammock, B. D., A highly sensitive and rapid ELISA for the arylurea herbicides diuron, monuron, and linuron, *J. Agric. Food Chem.*, 42, 413, 1994.

Schneider, P., Gee, S. J., Kreissig, S. B., Harris, A. S., Krämer, P., Marco, M. P., Lucas, A. D., and Hammock, B. D., Troubleshooting during the development and use of immunoassays for environmental monitoring, in *New Frontiers in Agrochemical Immunoassay,* Kurtz, D. A., Skerritt, J. H., and Stanker, L., Eds., AOAC, Arlington, VA, 1995, 103.

Siebert, S. T. A., Reeves, S. G., and Durst, R. A., Liposome immunomigration field assay device for alachlor determination, *Anal. Chim. Acta,* 282, 297, 1993.

Stanker, L. H., Watkins, B., Rogers, N., and Vanderlaan, M., Monoclonal antibodies for dioxin: antibody characterization and assay development, *Toxicology,* 45, 229, 1987.

Szurdoki, F., Bekheit, H. K. M., Marco, M. P., Goodrow, M. H., and Hammock, B. D., Synthesis of haptens and conjugates for an enzyme immunoassay for analysis of the herbicide bromacil, *J. Agric. Food Chem.,* 40, 1459, 1992.

Szurdoki, F., Kido, H., and Hammock, B. D., Development of assays for analysis of Hg(II) based on sulfur-containing ligands, in *Immunoanalysis of Agrochemicals: Emerging Technologies*, ACS Symposium Series 586, Nelson, J. O., Karu, A. E., and Wong, R. B., Eds., American Chemical Society, Washington, D.C., 1995, 248.

Tijssen, P., *Laboratory Techniques in Biochemistry and Molecular Biology, Volume 15: Practice and Theory of Enzyme Immunoassays*, Elsevier, Amsterdam, 1985.

Van Emon, J., Hammock, B. D., and Seiber, J. N., Enzyme-linked immunosorbent assay for paraquat and its application to exposure analysis, *Anal. Chem.*, 58, 1866, 1986.

Vanderlaan, M., Watkins, B. E., and Stanker, L., Environmental monitoring by immunoassay, *Environ. Sci. Technol.*, 22, 247, 1988.

Vo-Dinh, T., Tromberg, B. J., Griffin, G. D., Ambrose, K. R., Sepaniak, M. J., and Gardenhire, E. M., Antibody-based fiberoptics biosensor for the carcinogen benzo(a)pyrene, *Appl. Spectrosc.*, 41, 735, 1987.

Ward, V. K., Schneider, P., Kreissig, S. B., Karu, A., Hammock, B. D., and Choudary, P. V., Cloning, sequencing, and expression of the Fab fragment of a monoclonal antibody to the herbicide atrazine, *Protein Eng.*, 6, 981, 1993.

Ward, V. K., Hammock, B. D., Maeda, S., and Choudary, P. V., Development and application of recombinant antibodies to pesticide residue analysis, in *New Frontiers in Agrochemical Immunoassay*, Kurtz, D. A., Skerritt, J. H., and Stanker, L., Eds., AOAC, Arlington, VA, 1995a, 197.

Ward, V. K., Kreissig, S. B., Hammock, B. D., and Choudary, P. V., Generation of an expression library in the baculovirus expression vector system, *J. Virol. Meth.*, 53(2-3), 263, 1995b.

Wengatz, I., Schmid, R. D., Kreissig, S., Wittmann, C., Hock, B., Ingendoh, A., and Hillenkamp, F., Determination of the hapten density of immunoconjugates by matrix-assisted UV laser desorption/ionization mass spectrometry, *Anal. Lett.*, 25, 1983, 1992.

Wong, J. M., Li, Q. X., Hammock, B. D., and Seiber, J. N., Method for the analysis of 4-nitrophenol and parathion in soil using supercritical fluid extraction and immunoassay, *J. Agric. Food Chem.*, 39, 1802, 1991.

Wortberg, M., Cammann, K., Strupat, K., and Hillenkamp, F., A new non-enzymatic tracer for time-resolved fluoroimmunoassay of triazine herbicides, *Fres. J. Anal. Chem.*, 348, 240, 1994a.

Wortberg, M., Middendorf, C., Katerkamp, A., Rump, T., Krause, J., and Cammann, K., Flow-injection immunosensor for triazine herbicides using Eu(III)-chelate label fluorescence detection, *Anal. Chim. Acta*, 289(2), 177, 1994b.

Wylie, D. E., Carlson, L. D., Carlson, R., Wagner, F. W., and Schuster, S. M., Detection of mercuric ions in water by ELISA with a mercury-specific antibody, *Anal. Biochem.*, 194, 381, 1991.

Wylie, D. E., Lu, D., Carlson, L. D., Carlson, R., Babacan, K. F., Schuster, S. M., and Wagner, F. W., Monoclonal antibodies specific for mercuric ions, *Proc. Natl. Acad. Sci. U.S.A.*, 89, 4104, 1992.

Yalow, R. S. and Berson, S. A., Assay of plasma insulin in human subjects by immunological methods, *Nature*, 184, 1648, 1959.

15 Using Surface Probe Localized ^{31}P NMR Spectroscopy to Understand Sublethal Environmental Actions

Ronald S. Tjeerdema

INTRODUCTION

FUNDAMENTAL ENVIRONMENTAL TOXICOLOGY

The field of environmental toxicology revolves around investigating the interaction between toxic chemicals and target biological systems (Figure 15.1). Once released into the environment, chemicals may undergo a wide variety of natural processes, the sum of which is termed *environmental fate*. For instance, movement and transport within and between aquatic, atmospheric, lithospheric, and biotic environments may be highly influenced by a chemical's inherent molecular weight, water solubility (S), vapor pressure (P_v), octanol/water partition coefficient (K_{ow}), and soil sorption coefficient (K_d). It may dissolve into water, volatilize into the atmosphere, adsorb to soils or sediments, or bioconcentrate into organisms. In addition, chemical fate is highly dependent upon a wide variety of natural oxidative, reductive, hydrolytic, and photolytic reactions. The list of organisms ultimately targeted, as well as the degree of toxicity observed, is highly dependent upon the sum total of all the above processes, which must be clearly understood in order to accurately estimate chemical risk.

At the other end of the relationship are the organisms potentially targeted. Effects elicited by toxic chemicals may be either lethal or sublethal and may involve actions on physiological or biochemical processes. In general, chemicals are characterized by their most apparent toxic mechanism and site of action; however, in reality most chemicals probably act using multiple mechanisms at many sites. Whereas early investigations focused on the lethal effects of toxic chemicals, interest more recently has shifted to their inherent sublethal actions. A better understanding of such effects is important, as they may be of most influence on the current population vitality of many aquatic species. Presently, many populations are in decline; however, the presence of large numbers of dead adults (commonly referred to as a fish kill) is rarely observed. While other factors, such as habitat degradation, were thought to

1-56670-309-3/98/$0.00+$.50

ENVIRONMENTAL TOXICOLOGY

The Fundamental Process

FIGURE 15.1 Environmental toxicology is the organized investigation of the interactions of chemical agents, with their inherent environmental fate processes, and biological agents, with their inherent biochemical and physiological actions.

be the cause, toxic chemicals acting sublethally to stress adults, reducing growth or reproduction, may actually be responsible. However, such effects are difficult to measure.

Currently, measurement of the sublethal effects of toxic chemicals on aquatic organisms involves tests in which morphologic or behavioral responses are measured *in vivo*. They tend to be most useful for estimating harmful chemical concentrations (median-effect concentrations) and are rarely sensitive enough to detect sublethal biochemical actions. In contrast, description of biochemical actions generally has relied upon *in vitro* techniques, where neither whole organism responses nor the interactions of natural stress factors may be considered. Therefore, an approach suitable for measuring the sublethal biochemical actions of toxic chemicals in live, intact aquatic organisms under simulated environmental conditions is needed. *In vivo* nuclear magnetic resonance (NMR) spectroscopy can meet these requirements, because it can focus on biochemical and physiological processes in intact, non-stressed organisms.

In general, the NMR approach involves placement of a whole, healthy animal, contained within a flowing-water chamber, into an NMR magnet (Figure 15.2). Large intact adult animals such as abalones, oysters, or mussels require the use of both a wide-bore horizontal magnet and a surface probe which is placed adjacent to the organism to apply the required radio pulse and collect the resultant relaxation signal. There are a number of benefits in using such a localized approach. First, NMR allows measurement of low-energy nuclei, many of which are also naturally occurring, such as ^{1}H or ^{31}P. Second, it can focus on enriched nuclei, such as ^{13}C (which is commonly used to investigate drug biotransformation in rodents). As will be described, it also allows extensive work with intact individuals, facilitates the collection of kinetic information, may be made compatible with aquatic systems, and allows incorporation of environmental variables. Therefore, surface probe localized NMR can measure sublethal biochemical actions in intact aquatic organisms maintained in simulated environmental conditions.

Aquatic Organisms and Natural Stress Factors

Aquatic organisms, like all organisms, must routinely cope with changes in their physical environment. During the course of their existence, they may regularly

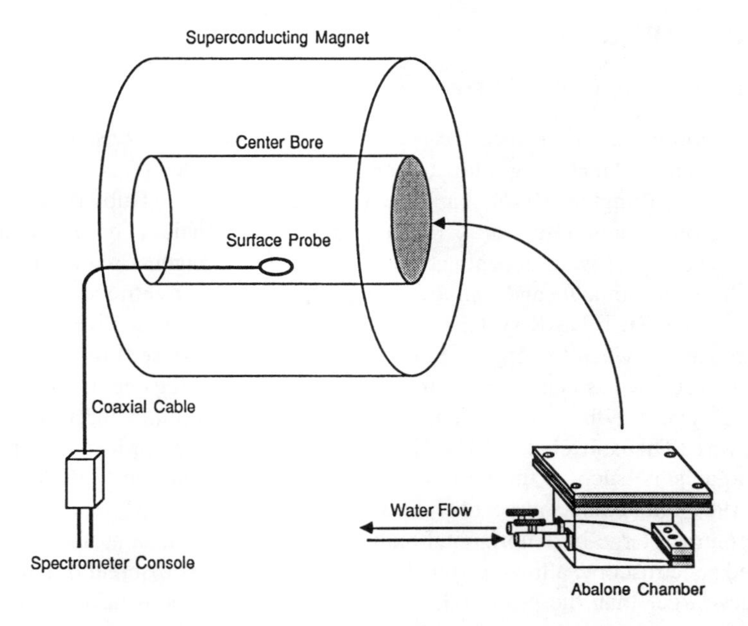

FIGURE 15.2 In using surface probe NMR to investigate sublethal effects in aquatic organisms, one places an intact individual in a flow-through chamber, which is then placed directly above the surface probe in a horizontal wide-bore NMR magnet.

experience changes in water temperature, pH, dissolved oxygen, salinity, and other dissolved salts, minerals, and nutrients. In general, they have developed many efficient ways to cope with such changes. The infusion of toxic chemicals into their physical environment simply represents another change that must be accommodated. However, in addressing the effects of toxic chemicals on aquatic organisms, rarely are interactions with other stress factors taken into account. Since organisms must adjust to many potential stress factors simultaneously, a research approach that facilitates measurement of such interactions will provide a superior estimate of the true environmental effects of toxic chemicals.

Development of the surface probe localized NMR approach for investigating sublethal effects first revolved around the use of the red abalone (*Haliotis rufescens;* Tjeerdema et al. 1991a). A marine gastropod mollusk of both commercial and environmental importance, it resides in lower intertidal and subtidal zones along most of the California coast. Since red abalones live in a transitional region between marine and terrestrial environments where both daily and seasonal variation is routine, they must regularly adjust to changes in numerous natural stress factors. For instance, seasonal variations in water temperature and salinity may range from 9 to 19°C and 17 to 34‰, respectively (Tjeerdema et al. 1991a, 1993). In addition, during a normal tidal cycle red abalones may be exposed to air for as long as 1 h (Tjeerdema et al. 1991b). Therefore, toxic chemicals may be viewed as simply another stress factor red abalones must routinely adjust to; how they interact with natural stresses must be understood in order to more accurately assess chemical risk. The surface probe localized NMR approach can provide such an assessment.

THE NMR APPROACH

Previous Aquatic NMR Research

Previous reports have described the use of *in vivo* NMR with excised tissues from aquatic organisms, including adductor muscles from cockles (Barrow et al. 1980) and mussels (Ellington 1983), barnacle depressor muscles (Dubyak and Scarpa 1983), abalone mantle (Burt et al. 1976), and deshelled snails (Thompson and Lee 1985). However, all were dependent upon narrow-bore magnets with conventional probes that accommodate and average signal acquisition over entire samples smaller than the probe. Such NMR systems are also useful for organ localization in whole organism studies when the organ can be isolated (but not excised) for insertion into the probe, such as has been done with crab leg muscles (Briggs et al. 1985) and the tail muscles of crayfish (Butler et al. 1985), shrimp (Kamp and Juretschke 1987), and prawns (Thebault et al. 1987). However, they are only applied to very small intact organisms, such as the cysts of brine shrimp (*Artemia* sp.; Drinkwater and Crowe 1987) and small mussels (*Mytilus edulis;* Aunaas et al. 1991).

In contrast, large-bore horizontal magnets can accommodate surface probes for localized spectroscopy, allowing signal acquisition from less discrete organs as well as tissues larger than the probe (Gadian 1982). Surface probes have been used to localize tissues in large intact organisms which are not discrete to conventional probes such as abalone foot (Higashi et al. 1989), fish lateral muscles (van den Thillart et al. 1989), and mussel adductor muscles (Fan et al. 1991). However, only recently has NMR of any type (surface probe localized), incorporating a unique aquatic exposure system, been used to investigate the toxic effects of chemicals in an aquatic organism such as the red abalone (Tjeerdema et al. 1991a, c, 1993).

Organism Suitability for NMR

In order to best utilize the surface probe localized NMR approach for investigating sublethal actions in large intact aquatic organisms, a number of important characteristics must be considered to facilitate signal acquisition. First, in order to observe environmentally realistic biochemical actions, organisms must not be anesthetized or physically restrained by any means that can introduce an unnatural stress. Therefore, organisms generally accustomed to remaining inactive for extensive time periods, such as echinoderms and mollusks, are preferred. Fishes may be used, but must be either restrained or anesthetized to prevent movement.

Organisms lacking obvious respiratory movements, again including echinoderms and mollusks, are also preferred, as the rhythmic beating of fish opercula may reduce NMR signal clarity. With fishes, signal clarity may be improved by using a surface probe of smaller diameter and placing it as far from the gills as possible, or by electronically gating the signal acquisition to match the opercular pumping rate.

For NMR investigations focused on energy metabolism, ^{31}P is the preferred nucleus. Thus, organisms must also possess organs or tissues rich in natural phosphorylated compounds, such as muscle. In order to optimally observe changes in natural metabolites (phosphagens), their cellular concentrations must be in the micromolar range. Surface probe localized NMR allows uncontaminated signal acquisition from discrete muscles. Therefore, the foot and shell muscles of gastropods, the adductor muscle of bivalves, and the lateral muscles of fishes are ideal subjects in that they provide both easy access to phosphagen-rich tissues and are required for life. For instance, if a toxic chemical were to impair the ability of an abalone's foot to properly contract, the animal could no longer attach tightly to surfaces, reducing its ability to graze and increasing its susceptibility to predators.

Therefore, for surface probe localized ^{31}P NMR, sessile invertebrates are preferred. In addition to the above criteria, intertidal marine species have the added benefit of occurring naturally in an environment of frequent change, and thus they are particularly useful for investigating the interactions of various natural stress factors with toxic chemicals. For instance, the red abalone was chosen for initial use not only because of its large, accessible foot muscle and sessile nature, but also because it naturally resides in intertidal and subtidal zones and thus is regularly subjected to dramatic changes in its environment.

The NMR Exposure System

Although a variety of flow-through exposure systems have previously been developed for use in toxicological studies, NMR presents unique demands not addressed by previous designs. The system developed for surface probe localized NMR consists of an exposure chamber and support system designed to meet a number of the requirements of intact aquatic organisms over long time periods in the magnet (Tjeerdema et al. 1991a). First, the system must provide a constant flow of fresh seawater, allowing for close control of both oxygenation and temperature. It must provide for reliable chemical dosing over extended time periods, and it must be relatively inert to toxic chemicals (to avoid chemical adsorption as well as plasticizer leaching). It must contain an aquatic organism within the magnet, near the surface probe, and discourage movement. The chamber and perfusion lines must be of nonmagnetic, nonconductive materials, and visually clear to assist probe placement. Also, the system must be of low intrinsic dead volume, because infusion lines must be long in order to keep the pumps away from the magnet and allow for electrical grounding and shielding. Finally, it must be rigorously leakproof to protect the surface probe and magnet from seawater corrosion.

The exposure chamber designed and developed for aquatic invertebrates and surface probe localized NMR is presented in Figure 15.3. It is an 11.7 (w) × 14.3 (l) × 6.2 (ht) cm box of 4.5-mm-thick clear acrylic polymer, except for the bottom surface (adjacent to the NMR surface probe), which is 1.5 mm thick. The top is sealed with a 4.5-mm-thick clear acrylic compression plate (15.9 × 18.4 cm), neoprene gasket, and nylon screws. Organisms are placed into a 2-mil modified fluoropolymer gas-sampling bag (12 × 15 cm) equipped with two fluoropolymer bulkhead compression tubing connectors (0.25-inch o.d.) anchored to one end of the box; the opposite end (open for organism insertion) is sealed with an acrylic clamp (2.2 × 10.5 cm), neoprene gasket, and nylon screws. The bag provides an inert primary containment for chemical exposure, while the box provides both a solid support and a secondary containment to ensure against leakage.

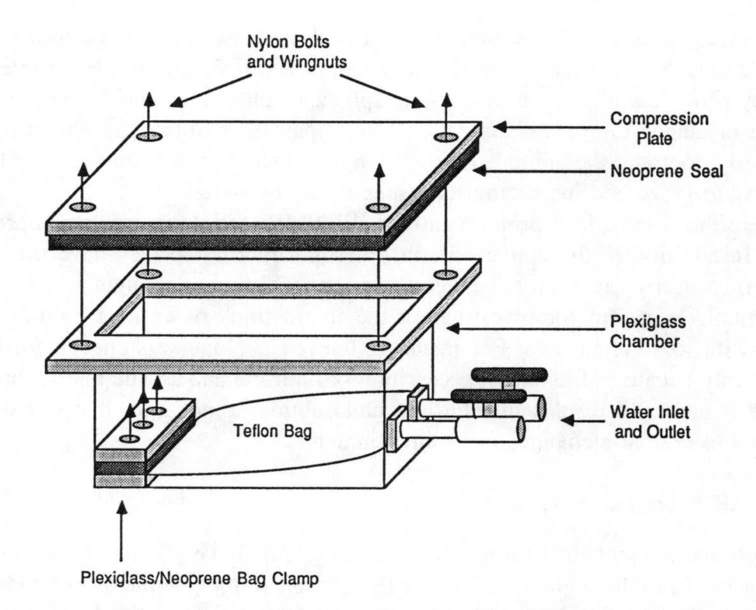

FIGURE 15.3 The fluoropolymer and acrylic exposure chamber for aquatic invertebrates. Water flow is restricted to the interior of the bag only, and the acrylic box serves as a solid support as well as a secondary water enclosure to protect the NMR magnet from leakage. (Adapted from Tjeerdema, R. S., Fan, T. W.-M., Higashi, R. M., and Crosby, D. G., *J. Biochem. Toxicol.*, 6, 45, 1991. Reprinted with permission from VCH Publishers © 1991.)

The support system is presented in Figure 15.4. Clean seawater is stored in a 20-L polypropylene carboy equipped with a spigot; temperature control is provided by a 25-ft coil of fluoropolymer-coated copper heat exchanger tubing (0.25-inch o.d.) connected to a recirculating water bath; and an aquarium air pump provides aeration. Water is pumped from the carboy with a peristaltic pump and silicone tubing (Cole-Parmer size 16), and all tube fittings are fluoropolymer compression-type (0.25-inch o.d.). A stock chemical solution is pumped into the water flow, using another pump, from a 6-L fluoropolymer modified gas-sampling bag (4-mil), which collapses upon emptying to eliminate headspace. The tubing leading to the chamber serves to mix the two flows, and for temperature control the bag is placed in a water bath. In total, the support system and exposure chamber more than adequately address all the demands considered in the design.

Information Derived from NMR

The ^{31}P nucleus is highly useful for *in vivo* NMR owing to its high natural abundance (100%), its presence in important endogenous "energy" compounds, and the use of inorganic monophosphate (P_i) as an intracellular pH (pH_i) marker (Gadian 1982). *In vivo* changes simultaneously measurable in intact organisms include fluctuations in concentrations of phosphagens such as phosphoarginine (PA, present in most invertebrates) or phosphocreatine (PC, present in most vertebrates), nucleoside phosphates

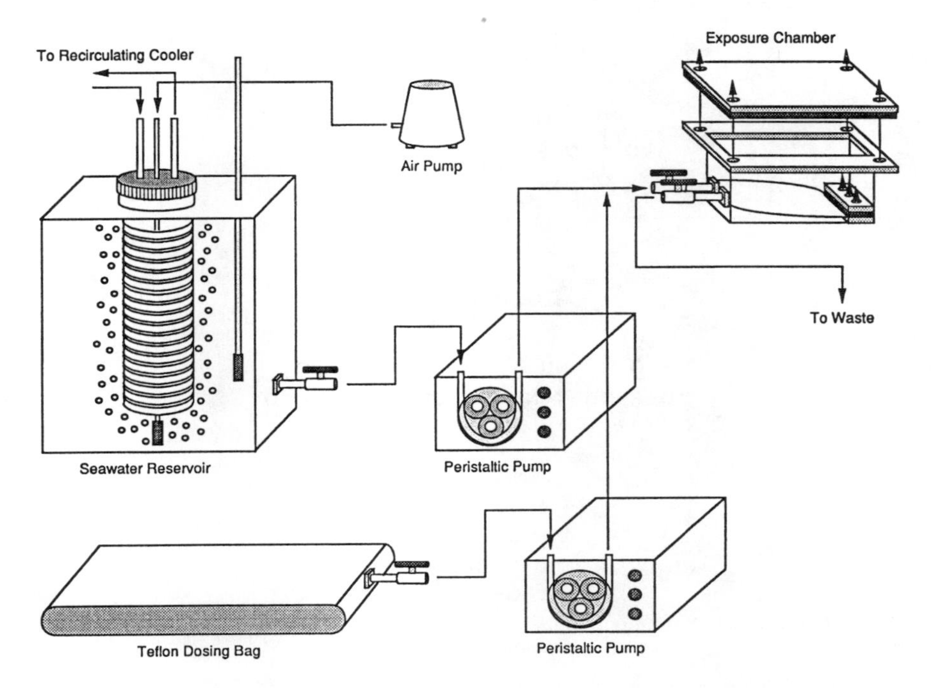

FIGURE 15.4 The flow-through exposure system for use with aquatic invertebrates and surface probe localized NMR. (Reprinted from Tjeerdema, R. S., Kauten, R. J., and Crosby, D. G., *Aquat. Toxicol.*, 21, 279, 1991, with kind permission from Elsevier Science Ltd, The Boulevard, Langford Lane, Kidlington 0X5 1GB, UK.)

(NPs, such as ATP), phosphoesters (e.g., glucose-6-phosphate), P_i, and pH_i (Figure 15.5; Gadian 1982).

In general, ^{31}P NMR spectra are acquired using a copper wire, 4-cm diameter, one-turn spiral surface probe placed in a General Electric CSI-2T® spectrometer equipped with a 2-Tesla 200-mm clear-bore horizontal magnet. Since exact quantitation of endogenous phosphagens by NMR is very difficult, a glass capillary tube filled with a solution of methylene diphosphonate (MDP) is mounted to the probe surface to serve as an external standard (Figure 15.5). For time-course studies, one-pulse data sets are typically acquired using the following conditions: a 50- to 55-μs pulse width, a ±1500-Hz spectral width, a 2-s pulse delay, and 2048 sample points (Tjeerdema et al. 1991b, c, 1993). Files average either 300 or 600 transients, which generally corresponds to 11.7 or 23.4 min of acquisition time, respectively.

A representative ^{31}P NMR spectrum for abalone foot muscle is presented in Figure 15.6. Spectral assignments are supported by previously published abalone spectra (Burt et al. 1976; Higashi et al. 1989) and spectra from standards of P_i, ATP, ADP, PC, PA, glucose-6-phosphate, $MgCl_2$, and NaCl (all at physiologic pH; Tjeerdema et al. 1991a). The NP peaks labeled D, E, and F are attributed to γATP + βADP, αATP + αADP + the phosphate moiety of nicotinamide adenine dinucleotide (NAD), and βATP, respectively (Figure 15.6; Tjeerdema et al. 1991a). Close similarity

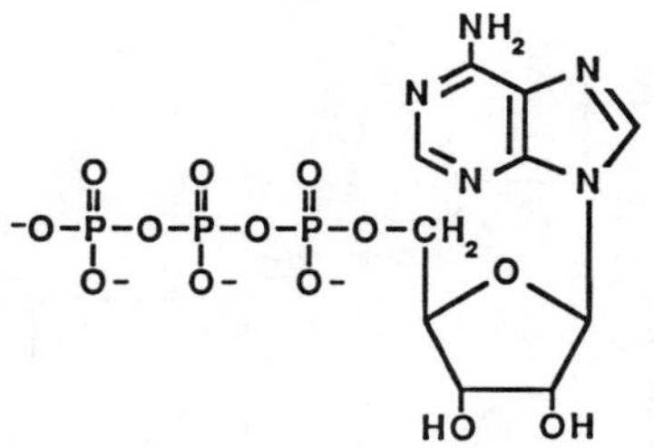

Adenosine Triphosphate (ATP)

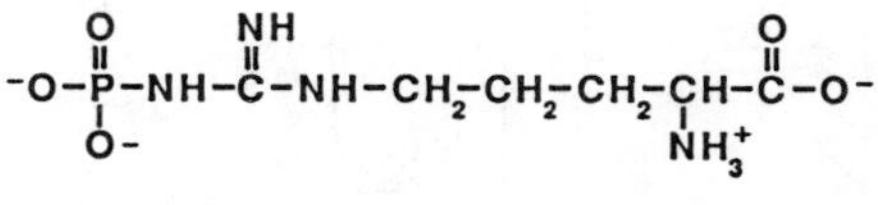

Phosphoarginine

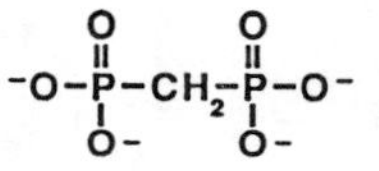

Methylene Diphosphonate

FIGURE 15.5 Structures of chemicals of importance for ^{31}P NMR, illustrating the unique molecular environments of their phosphates.

between *in vivo* and perchloric acid-extract spectra confirms that the *in vivo* spectrum represents only foot muscle (Higashi et al. 1989), and the measured longitudinal relaxation times (T_1 s) for PA (3.23 s), P_i (4.83 s), and the α- (1.12 s), β- (1.04 s), and γ- (1.55 s) NPs indicate partial saturation of resonances under the NMR conditions used, none of which influences phosphagen quantitation (Tjeerdema et al. 1991a).

In order to determine the *in vivo* kinetics of important endogenous phosphagens using NMR, relative changes in phosphagen signals may be measured by comparison to that of MDP, which remains constant (Tjeerdema et al. 1991a). In addition, changes in pH_i may be calculated from the difference in chemical shift between the P_i and PA resonances and a standard calibration curve (Tjeerdema et al. 1991a).

ACTIONS OF INDIVIDUAL STRESS FACTORS

PENTACHLOROPHENOL AND MITOCHONDRIAL FUNCTION

Pentachlorophenol (PCP) has been widely used as an antimicrobial wood preservative (Figure 15.7). Therefore, over the years it has been introduced into coastal waters from treated wood pilings and jetties, as well as from antifouling paints and drilling muds (Rao et al. 1979). It has been widely detected in aquatic animal tissues (Pierce and Victor 1978), and high concentrations in water have been associated with major fish kills (Pierce et al. 1977).

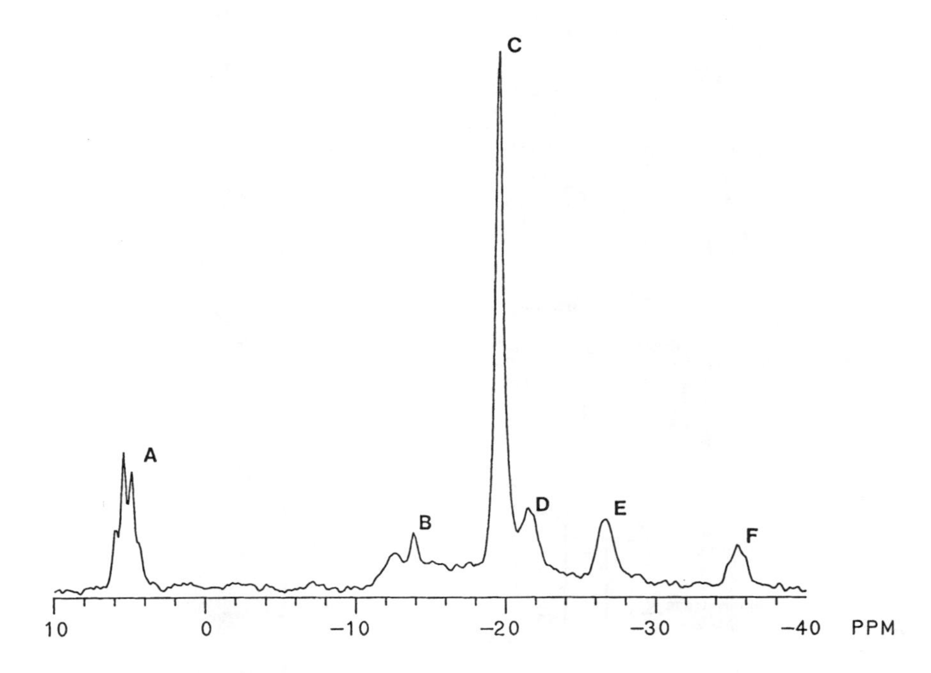

FIGURE 15.6 Representative ^{31}P NMR spectrum of nonstressed abalone foot muscle: A, methylene diphosphonate (external standard); B, inorganic phosphate; C, phosphoarginine; D, γ-phosphate of ATP with overlapping of the β-phosphate of ADP; E, α-phosphate of ATP with overlapping of the α-phosphate of ADP and the phosphate of nicotinamide adenine dinucleotide (NAD); and F, β-phosphate of ATP.

Similar to other phenols, PCP is thought to uncouple mitochondrial oxidative phosphorylation and thus disrupt ATP synthesis (Corbett et al. 1984). In general, transport of electrons down the cytochrome transport chain within the inner mitochondrial membrane causes protons to be pumped out through the membrane and against an electrochemical gradient; when they leak back in, they stimulate membrane-bound ATPases to synthesize ATP from ADP and P_i (Figure 15.7; Mitchell 1961, 1966; Mitchell and Moyle 1967). A dissociable acid (pK_a = 5.3), PCP may act by transporting protons back across the membrane to discharge the gradient and thus uncouple ATP synthesis from electron transport (Corbett et al. 1984).

Utilizing ^{31}P NMR to focus on endogenous phosphagens in aquatic invertebrates allows the monitoring of several very important metabolic reactions associated with ATP synthesis and use (Figure 15.7). When an abalone is unstressed, its foot muscle contains large concentrations of PA and ATP ([PA] and [ATP], respectively; ATP is best assessed by the β-NP peak); [P_i] may be nearly absent (Figure 15.6). However, when stressed, the most obvious changes are in [PA] and [P_i] (Figure 15.8). In general, unstressed abalones efficiently synthesize new ATP from ADP and P_i via mitochondrial oxidative phosphorylation, and transfer phosphate from ATP to arginine (catalyzed by arginine phosphokinase) to serve in an energy storage capacity (an

Cl Cl
Cl — O⁻ H⁺
Cl Cl

Pentachlorophenol

$$ATP + H_2O \rightleftharpoons ADP + H^+ + P_i$$

$$Phosphoarginine + ADP \rightleftharpoons ATP + Arginine$$

$$2\ ADP \rightleftharpoons ATP + AMP$$

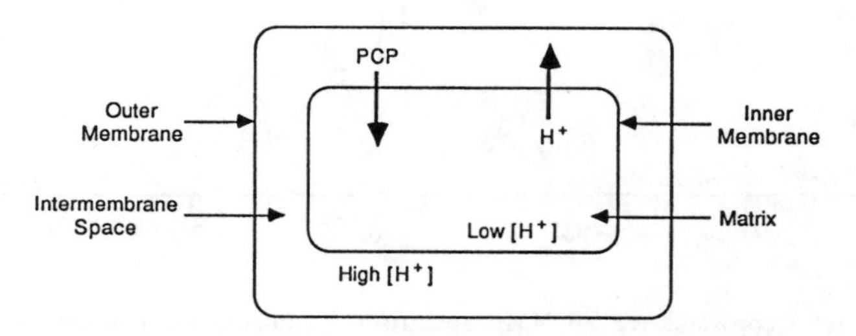

Highly Idealized Mitochondrion

FIGURE 15.7 The structure of pentachlorophenol (PCP; pK_a = 5.3) and the mechanism by which it uncouples mitochondrial oxidative phosphorylation. PCP is thought to discharge the proton gradient produced by electron transport within the inner mitochondrial membrane. The gradient normally powers membrane-bound ATPases, which produce ATP from ADP and inorganic phosphate (P_i). In the absence of stress, energy from ATP is transferred to arginine for storage. During uncoupling, ATP is regenerated for use from arginine phosphate.

energy savings account; Figure 15.7). When stressed, new ATP synthesis from oxidative phosphorylation is reduced, either via uncoupling or from the inhibition of electron transport which can occur when molecular oxygen, the terminal electron acceptor, is limited (such as during hypoxia). Therefore, the metabolic processes reverse: ATP is hydrolyzed to ADP and P_i during energy use, and because P_i cannot be recycled to synthesize new ATP, it rapidly accumulates. In order to meet metabolic demands, PA donates its phosphate to ADP to produce new ATP (also catalyzed by arginine phosphokinase; the savings account is drawn down); new ATP may also be formed from the conversion of two ADPs, which results in a decline of [ADP] over time (Figure 15.7). In essence, since the metabolic goal is to maintain near-constant [ATP], [PA] and [P_i] change most rapidly and thus represent some of the best endpoints measurable via *in vivo* ^{31}P NMR.

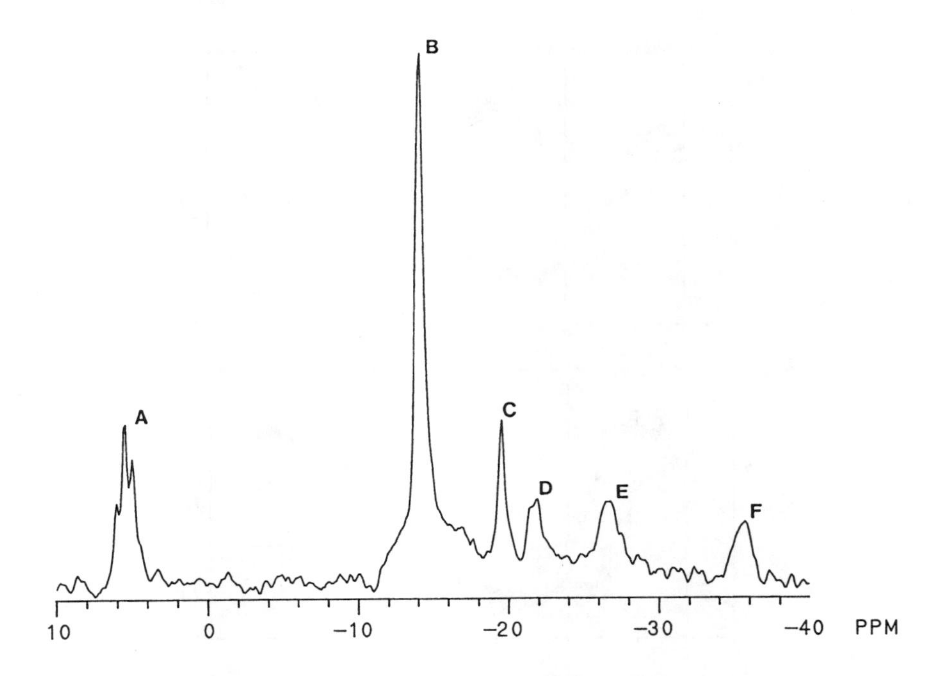

FIGURE 15.8 Representative ^{31}P NMR spectrum of pentachlorophenol-stressed abalone foot muscle, showing the accumulation of inorganic phosphate and depletion of phosphoarginine stores. (See Figure 15.6 for key.)

Sublethal Effects of Pentachlorophenol in Abalones

When individual red abalones are exposed to a sublethal concentration of PCP (1.2 mg L^{-1}) for 5 h and then allowed to recover for 13 h in clean seawater (14°C), a number of significant metabolic changes may be detected via surface probe localized ^{31}P NMR (Figure 15.9; Tjeerdema et al. 1991a). The most obvious changes involve those in foot muscle [PA] and [P_i] (Figure 15.9a). In all individuals observed, changes in both [PA] and [P_i] appeared complementary throughout both exposure and recovery; during PCP exposure, while [PA] initially increased by 20%, then declined to 80% of initial levels, [P_i] simultaneously decreased by 40%, then increased from 300 to 800% of preexposure levels. [PA] decline ceased within the first 2 h of recovery, then returned toward initial levels during the 13-h period. The increase in [P_i] also ceased during the initial 2 h, beyond which it returned to preexposure levels.

Foot muscle pH_i also changed during the exposure-recovery sequence (Figure 15.9b). Prior to exposure, pH_i ranged from 7.3 to 7.4. During the first 2 to 4 h of PCP exposure, it increased to between 7.4 and 7.5; however, in the later

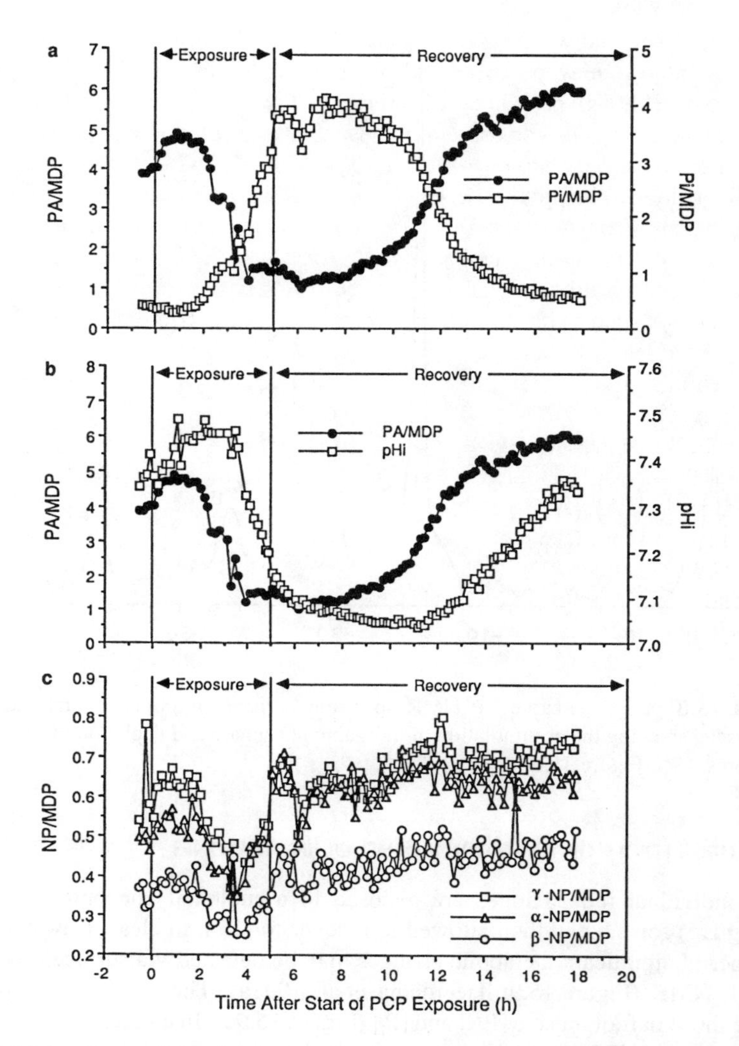

FIGURE 15.9 Comparison of the metabolic time-courses in a representative abalone during and after exposure to 1.2 mg L^{-1} of PCP at 14°C: (a) phosphoarginine (PA) and inorganic phosphate (P_i); (b) phosphoarginine (PA) and intracellular pH (pH_i); (c) the nucleoside phosphates (γ-NP, γ-phosphate of ATP with overlap of the β-phosphate of ADP; α-NP, α-phosphate of ATP with overlap of both the α-phosphate of ADP and the phosphate moiety of NAD; and β-NP, β-phosphate of ATP). Points indicate changes in peak intensities normalized to methylene diphosphonate (MDP), and each represents 512 spectra averaged over 9.63 min. (Adapted from Tjeerdema, R. S., Fan, T. W.-M., Higashi, R. M., and Crosby, D. G., *J. Biochem. Toxicol.,* 6, 45, 1991. Reprinted with permission from VCH Publishers © 1991.)

exposure and early recovery phases it significantly acidified. During recovery, pH_i eventually returned to at least, if not above, preexposure levels.

Surprisingly, [ATP] declined during the later part of PCP exposure; however, during recovery it returned to near preexposure levels (Figure 15.9c). From NMR peak assignments, and when the T_1s of the NPs are similar, the relative NP concentrations of interest may be expressed as follows: [ATP] = β, [ADP] = γ-β, and [NAD] = α-γ. Although peak intensities fluctuated, the fact that peak ratios remained steady indicated that any trends were mainly the result of changes in [ATP].

The use of surface probe localized ^{31}P NMR allows a number of sublethal metabolic actions to be monitored. However, NMR reveals only "free" metabolites; changes in peak intensities can also arise from changes in metabolite mobilization, as with ADP and P_i during ischemic hypoxia (Gadian 1982). Also, changes in the chemical shift of P_i, although normally attributed to changes in pH_i, may be affected by other factors, such as intracellular free ion composition (Gadian 1982). However, because of the lack of evidence to the contrary, and following convention, it may be assumed that changes in P_i chemical shift are primarily the result of those in pH_i (Tjeerdema et al. 1991a).

Overall, the metabolic effects of PCP coincided with its uptake and depuration profiles in abalones (Tjeerdema et al. 1991a; Tjeerdema and Crosby 1992). In addition, changes in [PA], [ATP], and pH_i were delayed by about 2 h, possibly reflecting the time required for PCP uptake and distribution to foot muscle. The inverse changes in [PA] and [P_i] induced by PCP exposure are consistent with both the blockage of mitochondrial ATP synthesis and [ATP] replenishment at the expense of [PA]. Also, changes in pH_i, although reflecting changes in energy phosphate metabolism, cannot be solely relied upon for metabolic interpretation. Whereas both glycolysis and ATP utilization result in acidification, conversion of PA to maintain [ATP] results in the opposite (Zubay 1988). Therefore, effects must first be characterized by changes in phosphate metabolites; concurrent changes in pH_i reflect both these and other effects (Tjeerdema et al. 1991a).

Considered an uncoupler of mitochondrial oxidative phosphorylation, PCP also displayed characteristics similar to those of the mitochondrial electron transport inhibitor 2-hydroxybiphenyl, which inhibits transport near coenzyme Q (ubiquinone; Oelze and Kamen 1975; Oelze et al. 1978). In addition, since [ATP] began to decline at approximately the same time as [PA], PCP may inhibit arginine phosphokinase and/or stimulate increased ATP utilization. In muscles from cockles (Barrow et al. 1980) and barnacles (Dubyak and Scarpa 1983) simultaneously exposed to inhibitors of both arginine phosphokinase and electron transport, [ATP] began decreasing before most PA was consumed; however, when barnacles were exposed to an electron transport inhibitor (cyanide) only, there was no difference in phosphagen response from that of hypoxia alone (pH_i was not reported; Dubyak and Scarpa 1983). Inhibition of arginine phosphokinase may account for the only partial recovery of [PA] during the recovery period.

Sublethal Effects of Hypoxia in Abalones

When individual red abalones are exposed to air for 1 h (to simulate low tide), then allowed to recover for at least 3 h in clean seawater (14°C), a number of significant metabolic changes may also be detected via surface probe localized ^{31}P NMR (Figure 15.10; Tjeerdema et al. 1991b). Again, as with PCP, the most obvious changes involved those in foot muscle [PA] and [P_i] (Figure 15.10a). Changes stimulated by hypoxia were complementary through exposure and recovery; during the

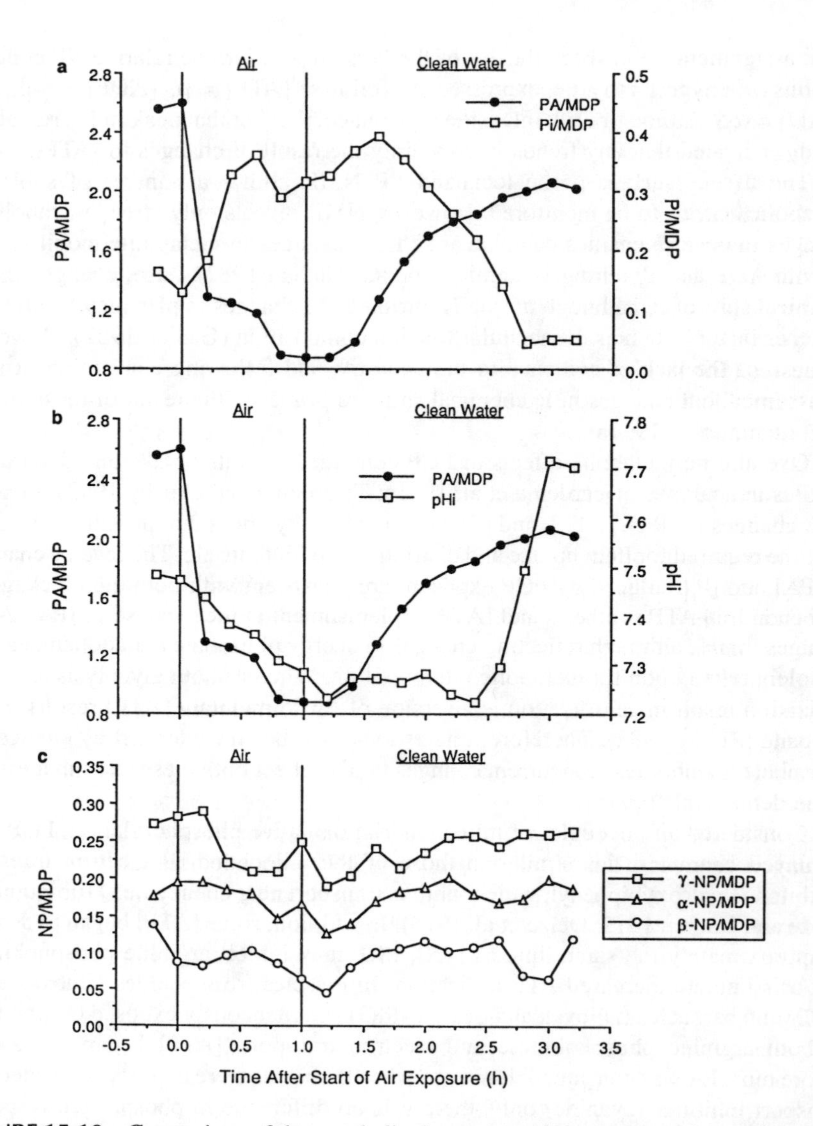

FIGURE 15.10 Comparison of the metabolic time-courses in a representative abalone during and after air exposure: (a) phosphoarginine (PA) and inorganic phosphate (P_i); (b) phosphoarginine (PA) and intracellular pH (pH_i); (c) the nucleoside phosphates (see Figure 15.9 for key). Points indicate changes in peak intensities normalized to methylene diphosphonate (MDP), and each represents 300 to 600 spectra averaged over 11.7 to 23.4 min. (Reprinted from Tjeerdema, R. S., Kauten, R. J., and Crosby, D. G., *Comp. Biochem. Physiol.*, 100B, 653-659, 1991, with kind permission from Elsevier Science Ltd, The Boulevard, Langford Lane, Kidlington 0X5 1GB, UK.)

rapid decline of [PA] to 70% of initial levels, [P_i] simultaneously increased by as much as 180%. Interestingly, the changes consistently stabilized during the later phase of air exposure. During resubmergence, [P_i] and [PA] generally recovered to prehypoxic levels within 4 h.

Foot muscle pH_i, which ranged from 7.5 to 7.6 prior to air exposure, declined rapidly with hypoxia to attain maximal acidification during the first hour of the clean water period (Figure 15.10b). During recovery, pH_i returned to near initial levels, although it generally lagged behind that of [PA]. Finally, unlike the decline observed during PCP exposure, no consistent change was observed in the NP resonances from the three abalones as a result of hypoxia.

The inverse changes in [PA] and [P_i] produced by hypoxia are consistent with the blockage of mitochondrial electron transport (from a deficiency in molecular oxygen to serve as the terminal electron acceptor; Zubay 1988). With ATP synthesis inhibited, [ATP] was replenished at the expense of [PA] to maintain basic life processes (Shoubridge and Radda 1984; Zubay 1988). The tissue acidification may result from both increased glycolysis (to increase basal ATP synthesis) and increased ATP utilization. The observed stabilization of [PA] and [P_i] prior to termination of air exposure may indicate the presence of an evolved compensatory mechanism to counteract the effects of hypoxia and assure survival during prolonged periods of emergence.

The metabolic effects of hypoxia were similar to those previously measured by NMR in excised tissues or intact organisms under either environmental or functional hypoxia. In muscles of cockles (*Tapes watlingi;* Barrow et al. 1980), barnacles (*Balanus nubilis;* Dubyak and Scarpa 1983), ribbed mussels (*Geukensia demissa;* Ellington 1983), bay mussels (*Mytilus edulis;* Ellington 1983; Fan et al. 1991), and crayfish (*Orconectes virilis;* Butler et al. 1985) exposed to hypoxic conditions, [PA], [P_i], and pH_i showed qualitatively similar responses to those in abalones. In muscles of carp (*Cyprinus carpio*) and goldfish (*Carassius auratus*), similar changes also occurred, except that PC replaced PA as the phosphagen (van den Thillart et al. 1989). The muscles of prawns (*Palaemon elegans;* Thebault et al. 1987) and shrimp (*Crangon crangon;* Kamp and Juretschke 1987) responded similarly to electrical stimulation, with a rapid decrease in [PA], increase in [P_i], and acidification of pH_i. However, [ATP] significantly decreased prior to complete depletion of [PA] (or [PC]) in only carp and goldfish, and the lack of measurable change in other species, including abalones, may indicate that arginine phosphokinase is fully functioning during hypoxia to catalyze the hydrolysis of PA, maintaining [ATP] and thus basic life processes (Tjeerdema et al. 1991b).

The metabolic effects of hypoxia in red abalones significantly differ from those previously measured by NMR in red and black abalones (*H. cracherodii*) exposed to hypersaline water (Higashi et al. 1989). While both [PA] and [P_i] responded by similarly decreasing and increasing, respectively, [ATP] also decreased and pH_i did not change; the decline in [ATP] may reflect an increased energy demand by ATP-dependent osmoregulatory processes. Because molecular oxygen was abundant, glycolysis, the tricarboxylic acid (TCA) cycle, and mitochondrial oxidative phosphorylation were presumably functioning, and thus no apparent acidification (from accumulation of lactate or TCA cycle intermediates) occurred. The effects also differed from those measured by NMR in red abalones exposed to PCP (Tjeerdema et al. 1991a).

Upon termination of exposure to either hypoxia, hypersalinity, or PCP, effects were generally reversed. The most rapid recovery occurred following hypoxia (generally within 3 h) and may indicate the rapidity with which molecular oxygen is replenished within muscle cells. The effects from hypersalinity generally required a longer recovery period (usually up to 5 h) and may reflect the time required for the return of tissues to near normal osmoticity (Higashi et al. 1989). Finally, the effects from PCP required the longest period to reverse (as much as 13 h) and may indicate the time required for depuration of the organic biocide (Tjeerdema et al. 1991a). Overall, the effects of hypoxia in the red abalone, as measured by ^{31}P NMR, indicate that although significant changes in both energy phosphagens and pH_i do occur, the changes stabilize during air exposure to allow survival during prolonged periods of tidal emergence.

INTERACTIONS OF MULTIPLE STRESS FACTORS

Some Potential Stress Interactions

Most traditional approaches to describing toxic effects involve the use of toxicity tests, where morphological, physiological, or behavioral endpoints are monitored during and after exposure to the toxicant. In general, such investigations incorporate a single variable (the test chemical) and hold all other important environmental parameters constant. However, in the real world, organisms are not neatly subjected to one stress at a time, but instead to a myriad of such factors, which may change constantly. For instance, abalones reside in upper subtidal and intertidal coastal regions. Therefore, they are potentially the target not only of toxic chemicals but also of changes in tides, temperature, and salinity, all of which may vary on daily and/or seasonal cycles. A toxic chemical may simply be viewed as an additional stress that must also be tolerated. Therefore, the most realistic approach to determining the actual impacts of toxic chemicals on organisms must consider multiple stress interactions. By allowing organisms to be maintained in simulated natural conditions, and accommodating changes in those conditions, NMR is an ideal tool for use in describing the biochemical actions of toxic chemicals in environmental conditions.

Pharmacologists have known for many years that multiple drug interactions pose serious consequences for patients undergoing multidrug therapy. Broadening the view, toxic chemicals can interact with natural stress factors in ways that may produce effects not predicted by traditional single-variable research approaches. In aquatic organisms, a number of possible water quality parameters, including temperature, salinity, pH, redox potential, and dissolved oxygen, may influence chemical toxicity in a number of different ways (Table 15.1). A toxic chemical and natural stress factor may interact in an additive way, where individual effects are summed, or in an antagonistic manner, where individual effects negate each other in combination. They can also interact in a synergistic manner, where the combined total of the actions is greater than their sum as individual effects, or potentiatively, in that a factor with no inherent activity enhances the effect of an active stress.

TABLE 15.1
Some Potential Environmental Interactions

	Type of Interaction	Mathematical Model
A.	Addition: The response elicited by a combined chemical and stress factor is *equal to* the combined responses of the two individual factors.	1 + 1 = 2
B.	Synergism: The response elicited by a combined chemical and stress factor is *greater than* the combined responses of the two individual factors.	1 + 1 = 3
C.	Potentiation: One factor which has no effect alone *enhances* the response from another factor.	1 + 0 = 2
D.	Antagonism: The response elicited by a combined chemical and stress factor is *less than* the combined responses of the two individual factors.	1 + 1 = 0

Interactions of Pentachlorophenol and Hypoxia in Abalones

Surface probe localized ^{31}P NMR has been used successfully to describe the interactive effects of PCP with hypoxia in red abalones as brought about by emergence during natural tidal fluctuations (Tjeerdema et al. 1991c). To do so, individual abalones were again exposed to 1.2 mg L^{-1} of PCP, but this time in combination with a 40-min emergence. A unique exposure sequence was designed to elucidate the nature of the interaction. In it, individual abalones were exposed to PCP within the NMR magnet until the spectral resonance of P_i was half-height to that of PA. This was done to assure that all replicate individuals were metabolically similar at the start of air exposure. They were allowed to stabilize for 2 h by exposure to clean flowing seawater (14°C; to stabilize the actions of PCP), exposed to air for 40 min, then resubmerged in clean flowing seawater for 15 h to check recovery. Stabilization provided a separation between the two stress factors so that the effects of PCP would not run onto those of hypoxia, confusing interpretation.

PCP-related effects were similar to those previously described (Tjeerdema et al. 1991a), as changes in [PA] and [P_i] were complementary through exposure and recovery, and both stabilized and began to reverse upon termination of dosing (prior to air exposure; Figure 15.11a). The effects of air exposure were additive to the effects of PCP; while [PA] rapidly declined to 60 to 90% of prehypoxic levels, [P_i] concurrently increased by up to 100%. The extent of response to hypoxia following PCP dosing was similar to that from hypoxia alone (Tjeerdema et al. 1991b). The decline in [PA] and increase in [P_i] both ceased on resubmergence, then recovered to initial levels within 3 and 1 h, respectively. Both [PA] and [P_i] fully recovered by the end of the 15-h clean seawater period.

Comparison of the time-courses of [PA] and pH_i reveals that PCP-related decreases in pH_i were also similar to those previously described (Tjeerdema et al.

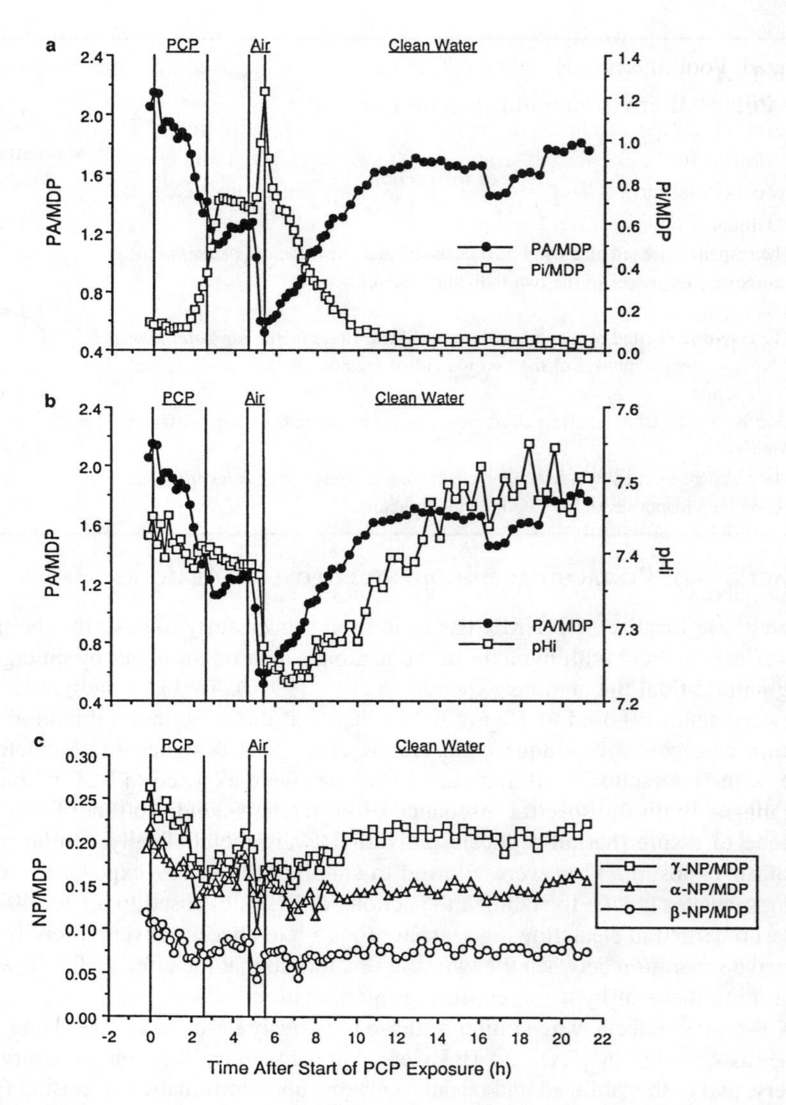

FIGURE 15.11 Comparison of the metabolic time-courses in a representative abalone during and after exposure to both 1.2 mg L^{-1} of PCP and air: (a) phosphoarginine (PA) and inorganic phosphate (Pi); (b) phosphoarginine (PA) and intracellular pH (pHi); (c) the nucleoside phosphates (see Figure 15.9 for key). Points indicate changes in peak intensities normalized to methylene diphosphonate (MDP), and each represents 300 to 600 spectra averaged over 11.7 to 23.4 min. (Reprinted from Tjeerdema, R. S., Kauten, R. J., and Crosby, D. G., *Aquat. Toxicol.*, 21, 279-294, 1991, with kind permission from Elsevier Science Ltd, The Boulevard, Langford Lane, Kidlington 0X5 1GB, UK.)

1991a; Figure 15.11b). Upon termination of dosing (and prior to air exposure), pH_i stabilized. Foot muscle pH_i averaged 7.4 prior to air exposure and declined rapidly with hypoxia to between 7.0 and 7.2 early into the clean water period. During recovery, it increased to prehypoxic levels within 14 h, with the total extent of change being similar to that observed from hypoxia alone (Tjeerdema et al. 1991b). Finally, all three NPs steadily declined during PCP exposure; however, no other changes were consistently observed as a result of hypoxia.

In general, the metabolic effects of hypoxia (in combination with PCP) were similar to those previously measured by NMR in excised tissues or intact organisms under environmental hypoxia alone (Barrow et al. 1980; Dubyak and Scarpa 1983; Ellington 1983; Butler et al. 1985; van den Thillart et al. 1989; Tjeerdema et al. 1991b). The deficiency in molecular oxygen, which serves as the terminal electron acceptor, effectively blocks mitochondrial electron transport (Tjeerdema et al. 1991b).

Exposure of red abalones to 1.2 mg L^{-1} of PCP also produced results similar to those previously described from PCP exposure alone (Tjeerdema et al. 1991a): [PA] declined and [P_i] concurrently increased, [ATP] declined prior to complete utilization of PA, and pH_i substantially decreased. Upon termination of dosing, all metabolic effects stabilized and began to reverse. Subsequent hypoxia caused rapid and concurrent decreases in both [PA] and pH_i and an increase in [P_i]. However, [ATP] did not continue to decline, and all effects were generally reversed within 2.5 h after termination of air exposure. The rapid reversal in the effects of hypoxia, even in combination with PCP exposure, indicates the rapid replenishment of molecular oxygen within muscle cells. The effects of PCP required a much longer time to reverse, possibly reflecting the time required for metabolic detoxication and/or depuration of the biocide. Overall, the effects of hypoxia in the presence of PCP were similar to those from hypoxia alone (they appeared superimposed upon those of PCP; Tjeerdema et al. 1991a). Thus, the effects may be considered to be additive.

Similar exposure of red abalones to 120 μg L^{-1} of PCP did not produce measurable effects, and effects of subsequent hypoxia were similar to those from hypoxia alone (Tjeerdema et al. 1991b, c). Thus, at environmental concentrations (California State Water Resources Control Board 1988), PCP does not significantly impact red abalones during short-term exposure, even with the additional stress imposed by tidal fluctuations.

Interactions of Pentachlorophenol and Temperature in Abalones

Surface probe localized ^{31}P NMR has also been used successfully to describe the interactive effects of PCP with another natural stress factor, water temperature, in red abalones as brought about by seasonal changes (Tjeerdema et al. 1993). To do so, individual abalones were again exposed to 1.2 mg L^{-1} of PCP, but this time at either 9 or 19°C, representing the seasonal water temperature range along the California coast. Again, individuals were exposed to PCP at one of the two temperatures within the NMR magnet until the spectral resonance of P_i was half-height to that of PA; this metabolic endpoint was selected for use in comparing the times required for onset of effects at the two temperatures. They were then exposed to clean flowing seawater for 16 h to check recovery.

At 9°C, PCP produced a significant decrease in [PA] and a concurrent increase in [P_i] (Figure 15.12a), an acidification of pH_i (from approximately 7.5 to 7.2; Figure 15.12b), and a decrease of up to 62% in [ATP] (again represented by the β-NP spectral resonance; Figure 15.12c). Onset of effects (when the area of the spectral P_i resonance was half that of PA) was attained in a mean time of 5.4 ± 0.5 h. On exposure to clean seawater, all metabolic effects generally reversed. After 10 h at 9°C, [PA] and [P_i] recovered by up to 75 and 53%, respectively (Figure 15.12a). Also, pH_i basified (from approximately 7.2 to 7.4; Figure 15.12b), and [ATP] exhibited a recovery of up to 77% (Figure 15.12c). Control abalones maintained at the low temperature alone did not exhibit adverse effects, indicating that low temperature alone does not cause measurable metabolic stress.

At 19°C, PCP-related effects were qualitatively similar to those at 9°C, as PCP produced a decrease of up to 67% in [PA] and an increase of up to 860% in [P_i] (Figure 15.13a), an acidification of pH_i (from 7.4 to 7.1; Figure 15.13b), and a decrease of up to 82% in [ATP] (Figure 15.13c). Average extent of response was greater at 19°C, and effects were produced more rapidly, as the metabolic endpoint was attained in less than half the time (2.4 ± 0.5 h; $p < 0.05$) as that at 9°C, indicating that the higher temperature significantly enhanced development of PCP-induced stress.

On exposure to clean seawater, all metabolic effects were generally reversed, as was observed at 9°C. After 10 h at 19°C, [PA] and [P_i] recovered by up to 94 and 92%, respectively (Figure 15.13a). Also, pH_i basified (from 7.2 to 7.4; Figure 15.13b), and [ATP] exhibited a recovery of up to 53% (Figure 15.13c). On average, the extent of recovery was greater at 19°C. Control abalones maintained at the high temperature alone did not exhibit adverse effects, indicating that high temperature alone does not cause measurable metabolic stress.

Exposure of red abalones to 1.2 mg L^{-1} of PCP at either 9 or 19°C produced effects in foot muscle qualitatively similar to those previously described from similar PCP exposure at 14°C (Tjeerdema et al. 1991a). Although neither low nor high water temperature alone produced measurable effects, increased water temperature served to hasten the onset of effects of the biocide; the time to onset of sublethal effects was significantly reduced by more than 50% in abalones exposed to PCP at 19 vs. 9°C ($p < 0.05$). Such an effect may reflect an increase in general metabolic activity in the abalone at higher temperatures, which thus enhances uptake of the biocide. Also, the average intensity of response to PCP and the average extent of recovery were both greater in individuals exposed to the biocide at the high temperature. The interaction of high water temperature with PCP is potentiative in nature; abalones maintained at 19°C alone exhibited no adverse effects, but exposure at that temperature enhanced the onset of sublethal metabolic effects.

Separate exposure of red abalones to 120 μg L^{-1} of PCP for 18 h at either low or high water temperature did not produce significant change; all metabolic endpoints

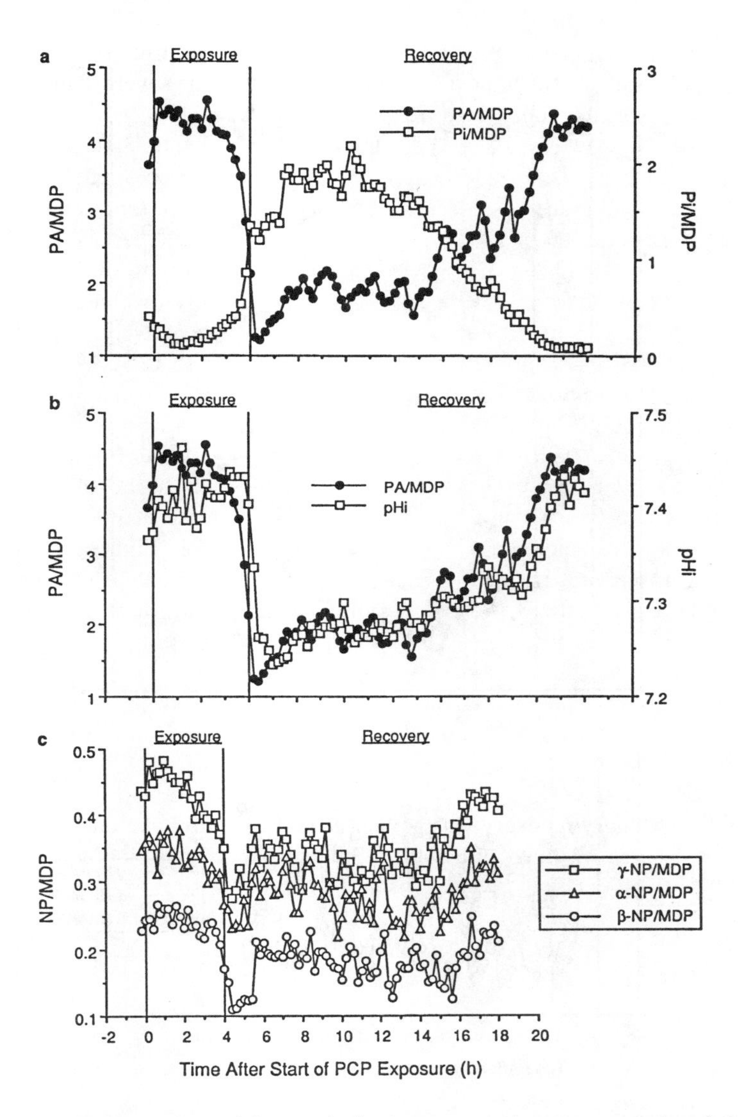

FIGURE 15.12 Comparison of the metabolic time-courses in a representative abalone during and after exposure to 1.2 mg L^{-1} of PCP at 9°C: (a) phosphoarginine (PA) and inorganic phosphate (P_i); (b) phosphoarginine (PA) and intracellular pH (pH_i); (c) the nucleoside phosphates (see Figure 15.9 for key). Points indicate changes in peak intensities normalized to methylene diphosphonate (MDP), and each represents 300 to 600 spectra averaged over 11.7 to 23.4 min. (Reprinted from Tjeerdema, R. S., Kauten, R. J., and Crosby, D. G., *Aquat. Toxicol.,* 26, 117, 1993, with kind permission from Elsevier Science Ltd, The Boulevard, Langford Lane, Kidlington 0X5 1GB, UK.)

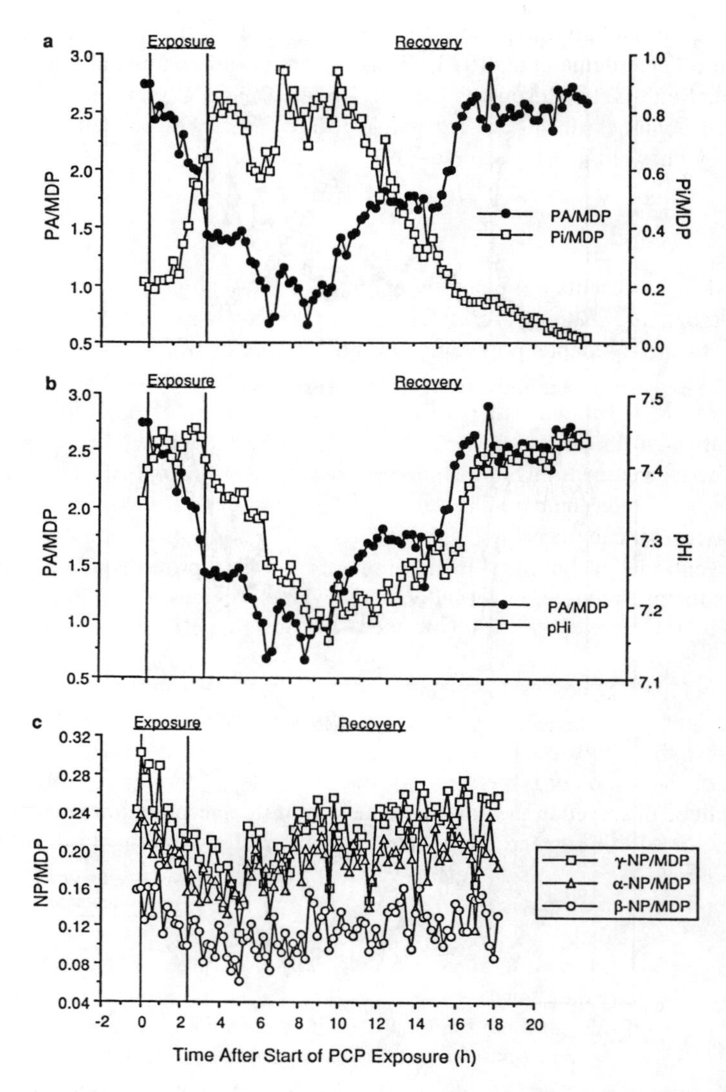

FIGURE 15.13 Comparison of the metabolic time-courses in a representative abalone during and after exposure to 1.2 mg L^{-1} of PCP at 19°C: (a) phosphoarginine (PA) and inorganic phosphate (P_i); (b) phosphoarginine (PA) and intracellular pH (pH_i); (c) the nucleoside phosphates (see Figure 15.9 for key). Points indicate changes in peak intensities normalized to methylene diphosphonate (MDP), and each represents 300 to 600 spectra averaged over 11.7 to 23.4 min. (Reprinted from Tjeerdema, R. S., Kauten, R. J., and Crosby, D. G., *Aquat. Toxicol.,* 26, 117, 1993, with kind permission from Elsevier Science Ltd, The Boulevard, Langford Lane, Kidlington 0X5 1GB, UK.)

remained stable and similar to those maintained in clean flowing seawater at each temperature (Tjeerdema et al. 1993). Thus, again, at environmental concentrations (California State Water Resources Control Board 1988), PCP does not significantly impact red abalones during short-term exposure, even with the additional stress imposed by changes in temperature.

CONCLUSIONS

In vivo NMR represents a major new approach for investigation of toxic effects in aquatic organisms. Using appropriate exposure systems and chambers, biochemical responses to environmental pollutants may now be measured in live, intact organisms as they occur (data are obtained in real time), allowing repetitive multicompound analyses to be performed using a single individual. The approach also facilitates normalization of all responses to an individual for greater interpretability of interactive effects, circumventing large individual variations. *In vivo* NMR also represents a more sensitive alternative to many toxicity measurements currently used in aquatic toxicology, and it can measure key parameters, such as pH_i, that cannot be measured with conventional techniques. In addition, the surface probe localized approach facilitates the use of large, intact individuals, which previously could not be accommodated by the narrow-bore magnets used to investigate effects in small organisms or excised tissues.

As a completely noninvasive technique, *in vivo* NMR can measure actual environmental responses at the biochemical level, providing important information on biochemical mechanisms of toxicity. Unlike with the use of isolated tissues, cell cultures, or individual enzyme systems, the information derived is very much like what might be observed in the environment, with little interpretation or extrapolation necessary. In addition, since the approach can model environmental conditions, it can incorporate changes in natural stress factors to provide an assessment of their interactions with toxic chemicals. Finally, *in vivo* NMR can accurately measure sublethal effects which may not kill adults outright, but which may reduce populations via decreased reproduction. Such information may be used more effectively than current toxicity data for the protection of aquatic populations from toxic chemicals. In essence, the combination of live, intact organisms, appropriate exposure chambers, and NMR can assist in determining molecular mechanisms of toxicity under a variety of realistic environmental conditions, providing a much more accurate assessment of the real environmental actions of toxic chemicals.

REFERENCES

Aunaas, T., Einarson, S., Southon, T. E., and Zachariassen, K. E., The effects of organic and inorganic pollutants on intracellular phosphate compounds in blue mussels (*Mytilus edulis*), *Comp. Biochem. Physiol.,* 100C, 89, 1991.

Barrow, K. D., Jamieson, D. D., and Norton, R. S., ^{31}P nuclear-magnetic-resonance studies of energy metabolism in tissue from the marine invertebrate *Tapes watlingi, Eur. J. Biochem.,* 103, 289, 1980.

Briggs, R. W., Radda, G. K., and Thulborn, K. R., ^{31}P-NMR saturation transfer study of the in vivo kinetics of arginine kinase in *Carcinus* crab leg muscle, *Biochim. Biophys. Acta,* 845, 343, 1985.

Burt, C. T., Glonek, T., and Bárány, M., Analysis of phosphate metabolites, the intracellular pH, and the state of adenosine triphosphate in intact muscle by phosphorus nuclear magnetic resonance, *J. Biol. Chem.,* 251, 2584, 1976.

Butler, K. W., Deslauriers, R., Geoffrion, Y., Storey, J. M., Storey, K. B., Smith, I. C. P., and Somorjai, R. L., ^{31}P nuclear magnetic resonance studies of crayfish (*Orconectes virilis*): the use of inversion spin transfer to monitor enzyme kinetics *in vivo, Eur. J. Biochem.,* 149, 79, 1985.

California State Water Resources Control Board, California State Mussel Watch: Ten-Year Data Summary, 1977–1987, Water Quality Monitoring Report 87-3, California State Water Resources Control Board, Sacramento, 1988.

Corbett, J. R., Wright, K., and Baillie, A. C., *The Biochemical Mode of Action of Pesticides,* 2d Ed., Academic Press, New York, 1984.

Drinkwater, L. E. and Crowe, J. H., Regulation of embryonic diapause in *Artemia:* environmental and physiological signals, *J. Exp. Zool.,* 241, 297, 1987.

Dubyak, G. R. and Scarpa, A., Phosphorus-31 nuclear magnetic resonance studies of single muscle cells isolated from barnacle depressor muscle, *Biochemistry,* 22, 3531, 1983.

Ellington, W. R., The extent of intracellular acidification during anoxia in the catch muscles of two bivalve mollusks, *J. Exp. Zool.,* 227, 313, 1983.

Fan, T. W.-M., Higashi, R. M., and Macdonald, J. M., Emergence and recovery of phosphate metabolites and intracellular pH in intact *Mytilus edulis* as examined *in situ* by *in vivo* ^{31}P- NMR, *Biochim. Biophys. Acta,* 1092, 39, 1991.

Gadian, D. G., *Nuclear Magnetic Resonance and Its Applications to Living Systems,* Oxford University Press, New York, 1982.

Higashi, R. M., Fan, T. W.-M., and Macdonald, J. M., Monitoring of metabolic responses of intact *Haliotis* (abalones) under salinity stress by ^{31}P surface probe localized NMR, *J. Exp. Zool.,* 249, 350, 1989.

Kamp, G. and Juretschke, H. P., An *in vivo* ^{31}P-NMR study of the possible regulation of glycogen phosphorylase *a* by phosphagen via phosphate in the abdominal muscle of the shrimp *Crangon crangon, Biochim. Biophys. Acta,* 929, 121, 1987.

Mitchell, P., Coupling of phosphorylation to electron and hydrogen transfer by a chemiosmotic type of mechanism, *Nature,* 191, 144, 1961.

Mitchell, P., Chemiosmotic coupling in oxidative and photosynthetic phosphorylation, *Biol. Rev.,* 41, 445, 1966.

Mitchell, P. and Moyle, J., Acid-base titration across the membrane system of rat liver mitochondria: catalysis by uncouplers, *Biochem. J.,* 104, 588, 1967.

Oelze, J. and Kamen, M. D., Separation of respiratory reactions in *Rhodospirillum rubrum:* inhibition studies with 2-hydroxybiphenyl, *Biochim. Biophys. Acta,* 387, 1, 1975.

Oelze, J., Fakoussa, R. M., and Hudewentz, J., On the significance of electron transport systems for growth of *Rhodospirillum rubrum, Arch. Microbiol.,* 118, 127, 1978.

Pierce, R. H., Jr., and Victor, D. M., The fate of pentachlorophenol in an aquatic ecosystem, in *Pentachlorophenol: Chemistry, Pharmacology, and Environmental Toxicology,* Rao, K. R., Ed., Plenum Press, New York, 1978, 41.

Pierce, R. H., Jr., Brent, C. R., Williams, H. P., and Reeves, S. G., Pentachlorophenol distribution in a freshwater ecosystem, *Bull. Environ. Contam. Toxicol.,* 18, 251, 1977.

Rao, K. R., Fox, F. R., Conklin, P. J., Cantelmo, A. C., and Brannon, A. C., Physiological and biochemical investigations of the toxicity of pentachlorophenol to crustaceans, in *Marine Pollution: Functional Responses,* Vernberg, W. B., Calabrese, A., Thurberg, F. P., and Vernberg, F. J., Eds., Academic Press, New York, 1979, 307.

Shoubridge, E. A. and Radda, G. K., A ^{31}P-nuclear magnetic resonance study of skeletal muscle metabolism in rats depleted of creatine with the analog β-guanidinopropionic acid, *Biochim. Biophys. Acta,* 805, 79, 1984.

Thebault, M. T., Raffin, J. P., and Le Gall, J. Y., *In vivo* ^{31}P NMR in crustacean muscles: fatigue and recovery in the tail musculature from the prawn *Palaemon elegans, Biochem. Biophys. Res. Comm.,* 145, 453, 1987.

Thompson, S. N. and Lee, R. W. K., ^{31}P NMR studies on adenylates and other phosphorus metabolites in the schistosome vector *Biomphalaria glabrata, J. Parasitol.,* 71, 652, 1985.

Tjeerdema, R. S. and Crosby, D. G., Disposition and biotransformation of pentachlorophenol in the red abalone (*Haliotis rufescens*), *Xenobiotica,* 22, 681, 1992.

Tjeerdema, R. S., Fan, T. W.-M., Higashi, R. M., and Crosby, D. G., Sublethal effects of pentachlorophenol in the abalone (*Haliotis rufescens*) as measured by *in vivo* ^{31}P NMR spectroscopy, *J. Biochem. Toxicol.,* 6, 45, 1991a.

Tjeerdema, R. S., Kauten, R. J., and Crosby, D. G., Sublethal effects of hypoxia in the abalone (*Haliotis rufescens*) as measured by *in vivo* ^{31}P NMR spectroscopy, *Comp. Biochem. Physiol.,* 100B, 653, 1991b.

Tjeerdema, R. S., Kauten, R. J., and Crosby, D. G., Interactive effects of pentachlorophenol and hypoxia in the abalone (*Haliotis rufescens*) as measured by *in vivo* ^{31}P NMR spectroscopy, *Aquat. Toxicol.,* 21, 279, 1991c.

Tjeerdema, R. S., Kauten, R. J., and Crosby, D. G., Interactive effects of pentachlorophenol and temperature in the abalone (*Haliotis rufescens*) as measured by *in vivo* ^{31}P-NMR spectroscopy, *Aquat. Toxicol.,* 26, 117, 1993.

van den Thillart, G., van Waarde, A., Muller, H. J., Erkelens, C., Addink, A., and Lugtenburg, J., Fish muscle energy metabolism measured by *in vivo* ^{31}P-NMR during anoxia and recovery, *Am. J. Physiol.,* 256, R922, 1989.

Zubay, G. L., *Biochemistry,* Macmillan, New York, 1988.

16 The Use of Developing Organisms for Assessing Contamination in the Marine Environment

Gary N. Cherr

INTRODUCTION

Embryos and larvae of marine organisms have been used for water quality assessments for many years (e.g., MacArthur 1924; Waterman 1937; Sugiyama 1950; Kobayashi 1971). These developing systems were often used to test the quality of sea waters (Wilson and Armstrong 1961) and for determining potential impacts of anthropogenic activities near urbanized regions (Okubo and Okubo 1962; Kobayashi et al. 1972; Woelke 1972). Echinoderm (primarily sea urchin and sand dollar) gametes and embryos have been favorite tools for many of these investigations. There is a wealth of knowledge available on the basic cellular and molecular mechanisms of development in these systems; much of our current knowledge of modern developmental biology is based on studies using the sea urchin (Giudice 1985; Leahy 1986).

There has been increased interest in developing marine organisms for investigations into mechanisms of xenobiotic perturbation (e.g., Baldwin et al. 1992; Garman et al. 1994a,b; Toomey and Epel 1993; Anderson et al. 1994; Sanders et al. 1994), biomonitoring of effluents and sediments (e.g., Williams et al. 1986; Cherr et al. 1987; Hunt and Anderson 1989; Long et al. 1990), and for identification of toxic constituents in complex effluents (Higashi et al. 1992; Cherr et al. 1993). Three groups of developing systems have been utilized for the majority of these studies: echinoderms, mollusks, and the brown algae. Recent advances in the application of these species to solving problems associated with contamination in the marine environment will be discussed in this chapter. While not intended to be a comprehensive review, the chapter will highlight key aspects of several of these systems which make them particularly well suited for studies on the cellular basis of ecotoxicology.

ECHINODERMS

Echinoderms are attractive marine systems for the study of xenobiotic action at cellular and biochemical levels. The sea urchin system is particularly well suited for

1-56670-309-3/98/$0.00+$.50

such studies in that: (1) gametes are readily obtained over a large portion of the year; (2) embryos are easily cultured, undergo synchronous development, and can be easily manipulated under laboratory conditions; (3) the biochemical and molecular events associated with select developmental events are very well characterized; (4) specific cell types can be isolated from the developing embryos which will differentiate under *in vitro* conditions; and, (5) specific macromolecules, which are developmentally regulated, have been isolated and can potentially be used as biochemical markers of xenobiotic perturbation.

The ability of sea urchin gametes to complete successful fertilization in the presence of contaminated water has been used as a toxicity assay for a number of years (Kobayashi 1970; Kobayashi 1984; Dinnel et al. 1987; Cherr et al. 1987). This assay involves exposing sperm to contaminants for 10 to 60 minutes and assessing their ability to fertilize eggs, which is easily detected microscopically by the presence of a fertilization envelope. Due to small volumes of sample (as low as 1 mL) and short periods of exposure (≤60 min), echinoderm sperm cell toxicity tests have received a great deal of attention for both toxicant identification studies (Higashi et al. 1992; Cherr et al. 1993) and biomonitoring (Dinnel et al. 1987; Anderson et al. 1994). Sea urchin embryonic development, to the pluteus stage, has also been used frequently for toxicological studies. The pluteus stage, which is characterized by the presence of an internal skeleton consisting of calcified spicules, is readily attained *in vitro* within 48 to 72 h (depending on the temperature), and the normal morphology is often perturbed by xenobiotics.

Studies suggest that the early pre-hatch embryo is not particularly susceptible to xenobiotics (Kobayashi 1980). However, by 16 to 20 h following fertilization, ciliated free-swimming embryos (blastula stage) are potentially exposed to a variety of environmental stressors (including xenobiotics), and it is starting at this stage that the embryos are capable of responding biochemically to them, such as in the expression of heat shock proteins (Giudice 1985). Gastrulation begins with ingression of primary mesenchyme cells (PMCs) into the blastocoelic cavity, and formation of the archenteron (primitive gut) is initiated. This is followed by fusion of PMCs to form a syncytial network that defines the geometry of the forming spicules of the embryonic skeleton (Decker and Lennarz 1988). Completion of spiculogenesis coincides with embryos attaining the pluteus larval morphology. Studies on the effects of xenobiotics on sea urchin development have indicated that gastrulation and spicule formation appear to be key events that are perturbed during early embryogenesis. For example, embryos exposed to oil production wastes exhibit abnormal archenteron formation and secretion of excess extracellular matrix material in the blastocoelic cavity; the latter correlates with the expression of a high-molecular-weight glycoprotein (Baldwin et al. 1992).

Skeletogenesis in sea urchin embryos is initiated by fusion of PMCs in the blastocoelic cavity, followed by intracellular calcium accumulation in the syncytium (Decker and Lennarz 1988). These events are dependent on metalloendoprotease activity (Roe et al. 1989) and the expression of a 130 kDa PMC membrane protein which is involved in intracellular calcium transport. It has been suggested that a calcium pump in the PMC plasma membrane plays a role in calcium homeostasis during skeletogenesis (Decker and Lennarz 1988). While these cellular targets are likely candidates for xenobiotic (e.g., metals) perturbation, they have not been used for such studies to date. Since PMCs can easily be isolated from embryos (Harkey and Whiteley 1980; Ettensohn and McClay 1987; Cherr et al. 1992) and spicule formation by isolated cells in culture is routinely accomplished, the key events associated with spiculogenesis (syncytia formation through fusion of PMCs and calcium accumulation) can be studied *in vitro*. Our preliminary studies have revealed that PMCs in culture can be employed to investigate the effects of metals (copper and barium) on cell aggregation/syncytia formation and spicule formation. Toxicological studies using embryonic stages of echinoderms, which take advantage of the amount of basic information available on developmental mechanisms, will likely lead to an understanding of the basis behind commonly observed teratogenic effects at the cellular and molecular levels in developing systems.

MOLLUSKS

Developing mollusk embryos have also been employed for many years for assessing water quality; for these studies, the organisms of choice have been bivalves, including oysters and mussels. Oyster embryo toxicity tests were developed in the Pacific Northwest in the late 1960s and early 1970s, and have been considered a standardized biomonitoring tool for a number of years (Woelke 1972). Oyster embryos have been shown to be particularly sensitive to metals and some pesticides (Davis 1961; Calabrese et al. 1973; MacInnes and Calabrese 1978). Embryos from the mussel *Mytilus* have also been used extensively for biomonitoring of the estuarine and marine environments (Armstrong and Milleman 1974; Beaumont et al. 1987). Most recently, embryos of the red abalone (*Haliotis rufescens*) have been used for toxicological monitoring (Hunt and Anderson 1989). While mollusk embryos have also been used in some studies for identifying toxicants in effluents (Higashi et al. 1992; Cherr et al. 1993), there is little information available regarding the mechanisms of teratogenic effects of xenobiotics.

The key developmental event which is used as an endpoint for most short-term mollusk embryo toxicity assays has been normal shell development. While some aspects of this have been refined (Cherr et al. 1990), little information is available regarding potential mechanisms of xenobiotic perturbation of molluskan development. This is probably due both to the complex nature of that development and the ease of focusing on larval shell development as a readily identifiable endpoint. While limited information is available on the effects of heavy metals on the cytoskeleton during cleavage and embryonic development in a gastropod (*Ilyanassa*) (Conrad 1988), little is known regarding cellular targets of xenobiotics in most molluskan embryos. In *Ilyanassa*, copper induces formation of a precocious polar lobe through induction of the formation of microfilaments, while silver does not allow the constricted band of microfilaments present to undergo developmentally regulated depolymerization (Conrad 1988).

A unique application of molluskan development in toxicology are the studies of Raimondi and Schmitt (1992) who conducted field studies using developing red abalone larvae. Since abalone larvae can be induced to undergo settlement and metamorphosis in the laboratory upon exposure to γ-aminobutyric acid (GABA), the ability of both laboratory- and field-exposed larvae to respond to GABA can readily be assessed. In this study, laboratory-reared larvae at different stages (pre-competent or competent on response to GABA) were transplanted in small cages to various sites near an ocean discharge of aqueous oil production waste for varying times, simulating potential exposure times based on currents in the region, and assessed for survivorship, settlement, and GABA response in the laboratory. In all cases where effects were observed, a spatial relationship with the discharge point was noted.

Only pre-competent larvae were affected with respect to settlement in the laboratory, suggesting that exposure to the oil production wastes alters settlement capabilities of abalone larvae. While the mechanism of this developmental perturbation is presently not known, this study is important in that it is the first demonstration of relevant biological effects to developing marine embryos in the water column. Furthermore, since the effects observed in this study were stage-specific during a realistic exposure regime in the field, the elucidation of stage-specific sensitivity of these, and other embryos, to a variety of stressors in the laboratory is of great importance. The definition of stage-specific effects and sensitivities is critical for ultimately understanding ecological impacts in coastal environments.

BROWN ALGAE

In addition to understanding the effects of contaminants on animal systems, there also has been interest in determining impacts in plants. Marine macroalgae play a key role in intertidal and subtidal communities, providing shelter and nursery habitat for invertebrates and fish, as well as a source of food for herbivores. Due to their importance and potential to be impacted by anthropogenic contaminants, marine macroalgae have recently become subjects of toxicological study (Steele and Hanisek 1979; Markham et al. 1980; Thursby and Steele 1984; Anderson et al. 1990). The giant kelp, *Macrocystis pyrifera,* is economically important and an essential part of California's coastal ecosystem. *M. pyrifera* has recently become an important test species for monitoring ocean discharges in California through the use of the microscopic gametophyte stage (which represents a portion of the reproductive phase of the life cycle) as a short-term chronic toxicity test for assessing pollution in the marine environment (Anderson and Hunt 1988; Anderson et al. 1990).

Our laboratory has investigated the cellular events associated with gametophyte development in *M. pyrifera* (Pillai et al. 1992). Reproduction in the giant kelp involves the release of haploid motile zoospores from the adult plant which settle onto a substrate, initiate germination within 4 to 5 h, and produce a 12 to 15-μm-long germ tube by 16 to 18 h. Shortly after completion of the germ tube growth, the zoospore nucleus undergoes a division, and one of the daughter nuclei is then immediately translocated to the distal end of the germ tube. Following this nuclear translocation, the first gametophytic cross wall is formed and, subsequently, the daughter nucleus remaining in the original zoospore body undergoes repositioning, assuming a position within the germ tube near the cross wall. Germ tube formation and nuclear translocation are temporally and mechanistically distinct (Pillai et al. 1992). During germination, actin dynamics are critical for normal germ tube growth, while microtubules are only involved in subsequent nuclear division and translocation

(Pillai et al. 1992). Furthermore, a number of rounds of DNA replication occur prior to the first nuclear division (Garman et al. 1994b). Thus, xenobiotics potentially have a number of cellular targets available in this developing system, and since these developmental events are temporally distinct, we have been able to assess the impacts of stage-specific exposures to contaminants (Garman et al. 1994a,b).

Nuclear events were found to be the most affected by copper, arsenic, and an aqueous oil-production waste (Garman et al. 1994a). Some contaminants are reversible in their inhibition, while others (e.g., arsenic) are not. Additional investigations focused on delineating the effects of arsenic (as arsenate) on DNA synthesis as compared to nuclear division and subsequent translocation (Garman et al. 1994b). Through discrete exposure regimes, both DNA synthesis and nuclear division/translocation were found to be inhibited by arsenate. Gametophytes exposed to arsenate at nuclear division did not appear to establish organized microtubule bundles (based on immunofluorescence microscopy) which surround the nuclei and traverse the germ tube during nuclear translocation (C. Vines, unpublished observation). Phosphate enrichment reduced the inhibitory effects of arsenate on division/translocation of the nucleus, supporting the hypothesis that arsenate interferes with phosphorylation. However, this reduction in arsenate inhibition may also be due to phosphate competition of arsenate uptake by the developing gametophyte.

By exposing *M. pyrifera* gametophytes to toxicants at times which correspond to key developmental events, we have been able to demonstrate that environmentally relevant levels of arsenic are capable of disrupting microtubule functioning at the time of nuclear division. However, the cell cycle also appears to be disrupted by this metal since DNA synthesis is inhibited. While gametophytes exposed to lower concentrations of select toxicants may undergo normal germination and germ tube formation, they can be developmentally arrested with respect to the cell cycle and/or nuclear events; thus, subsequent normal development will not proceed.

CONCLUSIONS

The advantage of utilizing some of the developing systems discussed above is that many of the cellular and molecular events associated with animal development have been investigated using echinoderms, while brown algal development is a model system commonly employed in plant development (reviewed by Callow et al. 1985; Kropf 1992). The sea urchin system has been routinely utilized as a general model system for cellular, biochemical, and molecular events in developing animal cells; as such, information on perturbation at these levels can be extrapolated to other systems. Finally, sea urchin fertilization and development, molluskan development, and giant kelp gametophyte development have all been utilized in applied toxicological studies and are currently in use for regulatory purposes (U.S. EPA 1988; California Ocean Plan 1990). Thus the opportunity to obtain mechanistic information for select toxicants using these systems has never been as great as at the present. It is expected that the utilization of these developing systems as tools for assessing the quality of our coastal waters, while at the same time attaining an understanding of the cellular and subcellular targets of contaminants, will be important factors in the future of marine toxicology.

REFERENCES

Anderson, B. S. and Hunt, J. W., Bioassay methods for evaluating toxicity using *Macrocystis pyrifera*., *Mar. Environ. Res.*, 26, 113, 1988.

Anderson, B. S., Hunt, J. W., Turpen, S. L., Coulon, A. R., and Martin, M., Copper toxicity to microscopic stages of giant kelp *Macrocystis pyrifera*: interpopulation comparisons and temporal variability, *Mar. Ecol. Prog. Ser.*, 68, 147, 1990.

Anderson, S. L., Hose, J. E., and Knezovich, J., Genotoxic and developmental effects in sea urchins are sensitive indicators of effects of genotoxic chemicals, *Environ. Toxicol. Chem*, 13, 1033, 1994.

Armstrong, D. A. and Millemann, R. E., Effects of the insecticide Sevin and its first hydrolytic product 1-naphthol on some early developmental stages of the bay mussel, *Mytilus edulis*, *Mar. Biol.*, 28, 11, 1974.

Baldwin, J. D., Pillai, M. C., and Cherr, G. N., Response of embryos of the sea urchin *Strongylocentrotus purpuratus* to aqueous petroleum waste includes the expression of a high molecular weight glycoprotein, *Mar. Biol.*, 114, 21, 1992.

Beaumont, A. R., Tserpes, G., and Budd, M. D., Some effects of copper on the veliger larvae of the mussel *Mytilus edulis* and the scallop *Pecten maximus*, *Mar. Environ. Res.*, 21, 299, 1987.

Calabrese, A., Collier, R. S., Nelson, D. A., and MacInnes, J. R., The toxicity of heavy metals to embryos of the American oyster *Crassostra virginica*, *Mar. Biol.*, 18, 162, 1973.

California Ocean Plan, *Functional Equivalent Document*, State Water Resources Control Board, 1990, 105 pp.

Callow, J. A., Callow, M. E., and Evans, L. V., Fertilization in *Fucus*, in *Biology of Fertilization*, Metz, C. B. and Monroy, A., Eds., Vol. 2, Academic Press, Orlando, 1985, 389.

Cherr, G. N., Shenker, J. M., Lundmark, C., and Turner, K. O., Toxic effects of selected bleached kraft mill effluent constituents on the sea urchin sperm cell, *Environ. Toxicol. Chem.*, 6, 561, 1987.

Cherr, G. N., Shoffner-McGee, J., and Shenker, J. M., Methods for assessing fertilization and embryonic/larval development in toxicity tests using the California mussel (*Mytilus californianus*), *Environ. Toxicol. Chem.*, 9, 1137, 1990.

Cherr, G. N., Summers, R. G., Baldwin, J. D., and Morrill, J. B., Preservation and visualization of the sea urchin blastoceolic extracellular matrix, *Microsc. Res. Tech.*, 22, 11, 1992.

Cherr, G. N., Fan, T. W-M., Pillai, M. C., Shields, T., and Higashi, R. M., Electrophoretic separation, characterization, and quantification of biologically active lignin-derived macromolecules, *Anal. Biochem.*, 214, 521, 1993.

Conrad, G. W., Heavy metal effects on cellular shape changes, cleavage, and larval development of the marine gastropod mollusk, *Ilyanassa obsoleta* Say., *Bull. Environ. Contam. Toxicol.*, 41, 79, 1988.

Davis, H. C., Effects of some pesticides on eggs and larvae of oysters (*Crassostrea virginica*) and clams (*Venus mercenaria*), *Comml. Fish. Rev.*, 23, 8, 1961.

Decker, G. L. and Lennarz, W. J., Skeletogenesis in the sea urchin embryo, *Development*, 103, 231, 1988.

Dinnel, P. A., Link, J. M., and Stober, Q. J., Improved methodology for a sea urchin sperm cell bioassay for marine waters, *Arch. Environ. Contam. Toxicol.*, 16, 23, 1987.

Ettensohn, C. A. and McClay, D. R., A new method for isolating primary mesenchyme cells of the sea urchin embryo, *Exptl. Cell Res.*, 168, 431, 1987.

Garman, G. D., Pillai, M. C., and Cherr, G. N., Inhibition of cellular events during algal gametophyte development: effects of select metals and an aqueous petroleum waste, *Aquatic Toxicol.*, 28, 127, 1994a.

Garman, G. D., Pillai, M. C., Goff, L. J., and Cherr, G. N., Nuclear events during early development in *Macrocystis pyrifera* gametophytes and the temporal effects of a marine contaminant, *Mar. Biol.,* 121, 355, 1994b.

Giudice, G., *The Sea Urchin Embryo: A Developmental Biological System*, Springer Verlag, Berlin, 1985.

Harkey, M. A. and Whiteley, A. H., Isolation, culture, and differentiation of echinoid primary mesenchyme cells, *Wilhelm Roux Arch.,* 189, 111, 1980.

Higashi, R. M., Cherr, G. N., Shenker, J. M., Macdonald, J. M., and Crosby, D. G., A polar high molecular mass constituent of bleached kraft mill effluent is toxic to marine organisms, *Environ. Sci. Tech.,* 26, 2413, 1992.

Hunt, J. W. and Anderson, B. S., Sublethal effects of zinc and municipal effluents on larvae of the red abalone *Haliotis rufescens, Mar. Biol.,* 101, 545, 1989.

Kobayashi, N., Bioassay data for marine pollution using sea urchin eggs, *Publ. Seto Mar. Biol. Lab.,* 18, 421, 1970.

Kobayashi, N., Fertilized sea urchin eggs as an indicatory material for marine pollution bioassay, preliminary experiments, *Publ. Seto Mar. Biol. Lab.,* 18, 376, 1971.

Kobayashi, N., Nogami, H., and Doi, K., Marine pollution bioassay by using sea urchin eggs in the Inland Sea of Japan (The Seto-Naikai), *Publ. Seto Mar. Biol. Lab.,* 19, 359, 1972.

Kobayashi, N., Comparative sensitivity of various developmental stages of sea urchins to some chemicals, *Mar. Biol.,* 58, 163, 1980.

Kobayashi, N., Marine ecotoxicological testing with echinoderms, in *Ecotoxicological Testing for the Marine Environment,* Persoone, G., Jaspers, E., and Claus, C., Eds., State Univ. Ghent and Inst. Mar. Scient. Res., Bredene, Belgium, Vol. 1, 1984, 341.

Kropf, D. L., Establishment and expression of cellular polarity in fucoid eggs, *Microbiol Rev.,* 56, 316, 1992.

Leahy, P. S., Laboratory culture of *Strongylocentrotus purpuratus* adults, embryos, and larvae, *Meth. Cell Biol.,* 27, 1, 1986.

Long, E. R., Buchman, M. F., Bay, S. M., Breteler, R. J., Carr, R. S., Chapman, P. M., Hose, J. E., Lissner, A. L., Scott, J., and Wolfe, D. A., Comparative evaluation of five toxicity tests with sediments from San Francisco Bay and Tomales Bay, California, *Env. Toxicol. Chem.,* 9, 1193, 1990.

MacArthur, J. W., An experimental study and a physiological interpretation of exogastrulation and related modifications in echinoderm embryos, *Biol. Bull.,* 46, 60, 1924.

MacInnes, J. R. and Calabrese, A., Response of the embryos of the American oyster, *Crassostrea virginica,* to heavy metals at different temperatures, in *Behavior of Marine Organisms,* Pergamon Press, Elmsford, 1978, 195.

Markham, J. W., Kremer, B. P., and Sperling, K. R., Effects of cadmium on *Laminaria saccharina* in culture, *Mar. Ecol. Prog. Ser.,* 3, 31, 1980.

Nacci, D. E., Jackim, and Walsh, R., Comparative evaluation of three rapid marine toxicity tests: sea urchin early embryo growth test, sea urchin sperm cell toxicity test, and Microtox, *Env. Toxicol. Chem.,* 5, 521, 1986.

Okubo, K. and Okubo, T., Study on the bio-assay method for the evaluation of water pollution — II. Use of the fertilized eggs of sea urchins and bivalves, *Bull. Tokai Reg. Fish. Res. Lab.,* 32, 131, 1962.

Pillai, M. C., Baldwin, J. D., and Cherr, G. N., Early development in an algal gametophyte: regulation of germination and nuclear events by cytoskeletal elements, *Protoplasma,* 170, 34, 1992.

Raimondi, P. R., and Schmitt, R. J., Effects of produced water on settlement of larvae: field tests using red abalone, in *Produced Water: Technological/Environmental Issues and Solutions,* Ray J. P. and Engelhardt, F. R., Eds., Plenum Press, New York, 1992, 415.

Roe, J. L., Park, H. R., Strittmatter, W. J., and Lennarz, W. J., Inhibitors of metalloendoproteases block spiculogenesis in sea urchin primary mesenchyme cells, *Exptl. Cell Res.*, 181, 542, 1989.

Sanders, B. M., Martin, L. S., Nakagawa, P. A., Hunter, D. A., Miller, S., and Ullrich, S. J., Specific cross-reactivity of antibodies raised against two major stress proteins, stress 70 and chaperonin 60, in diverse species, *Environ. Toxicol. Chem.*, 13, 1241, 1994.

Steele, R. L. and Hanisak, M. D., Sensitivity of some brown algal reproductive stages to oil pollution, *Proc. Intl. Seaweed Symp.*, 9, 181, 1979.

Sugiyama, M., Polyspermy in sea urchin eggs induced by $CuSO_4$, *Zool. Magz.*, 50, 11, 1950.

Toomey, B. H. and Epel, D., Multixenobiotic resistance in *Urechis caupo* embryos: protection from environmental toxins, *Biol. Bull.*, 185, 355, 1993.

Thursby, G. B. and Steele, R. L., Toxicity of arsenite and arsenate to the marine microalga *Champia parvula* (Rhodophyta), *Environ. Toxicol. Chem.*, 3, 391, 1984.

United States Environmental Protection Agency, Short-term methods for estimating the chronic toxicity of effluents and receiving waters to marine and estuarine organisms, EPA/600/4-87/028, Washington, D.C., 1988.

Waterman, A., Effect of salts of heavy metals on development of the sea urchin, *Arbacia punctulata*, *Biol. Bull.*, 73, 401, 1937.

Williams, L. G., Chapman, P. M., and Ginn, T. C., A comparative evaluation of marine sediment toxicity using bacterial luminescence, oyster embryo, and amphipod sediment bioassays, *Mar. Environ. Res.*, 19, 225, 1986.

Wilson, D. P. and Armstrong, F. A. J., Biological differences between sea waters: experiments in 1960, *J. Mar. Biol. Assn. U.K.*, 41, 663, 1961.

Woelke, C. E., Development of a receiving water quality bioassay criterion based on the 48 hour Pacific oyster (*Crassostrea gigas*) embryo, Washington Dept. Fisheries, Tech. Rep. No. 9, 93 pp, 1972.

Epilogue: Multiple Stresses in Ecosystems

Donald G. Crosby

It seems to have become a tradition for the chairman of a conference or editor of a multiauthored work to summarize the results as a final penance. The thoughtful conclusions of the individual chapters need not be repeated here, but let me review instead some especially strong personal impressions from them.

The subject is multiple stresses in ecosystems, with special attention to environmental contaminants. Although often dominating our considerations, contaminants do not operate in the absence of other stressors such as climate, disease, and additional chemical factors such as water quality which may be of major dimensions. The contaminant picture itself is more complex than one might have expected only a few years ago, involving such subtle but important chemical and physical characteristics as volatilization, photodegradation, and microbial transformation. Indeed, the original definition of ecotoxicology, provided by Renee Truhaut in 1974, referred to it as the science concerned with the movement, fate, and biological effects of toxic chemicals ***in an ecosystem context***, a concept only now being addressed by volumes such as this.

As pointed out by **S. A. Levin**, we actually know far more about the fate and transport of contaminants than we do about their impacts, although the principles that underlie the environmental movement and degradation of chemicals are themselves important to an understanding of ecosystem processes. The key problem often seems to be one of scale: Can ecosystems really be treated in the same way as individuals in time and space, as is often assumed today? Although this issue provided a central focus of subsequent discussions, there seems to be agreement that, for purposes of experiment, perhaps an ecosystem might be viewed as a collection of smaller associated units, as Dr. Levin skillfully demonstrated. **S. M. Adams** also incorporated this idea into his chapter with a dual-axis scale relating time interval to toxicological and ecological relevance, the responses represented as "biomarkers."

There can be no doubt that multiple ecosystem stresses do exist and have demonstrable impacts. For example, the effects of human activities on Lake Tahoe's biological productivity as measured by loss of clarity described by **C. R. Goldman**, their effects on aquatic life in the Sacramento River and delta as outlined by **A. G. Heath**, effects on the marine environment cited by **E. D. Goldberg**, and the damage done to plants by atmospheric pollutants discussed by **G. E. Taylor, Jr.**, provided all the evidence one would ever need. However, seen through **D. W. Anderson**'s perspective, "everything already had undergone change" before human arrival, and so even the modern chemical stressors appear to be nothing new to ecosystems, albeit their potency and persistence often may be unprecedented. As Dr. Goldman correctly pointed out, it remains for present-day science to "tease out" the most important stress variables as they now exist in order to anticipate future human impacts, that is, just how ecosystem health will be affected.

This term "health" employed by **B. W. Wilson** and others must be used advisedly here, as it continues an argument that amateur ecologists such as myself should not touch with Dr. Mineau's proverbial 10-foot pole. The viewpoint that ecosystems simply can't be ascribed the characteristics of individuals, or even of populations and communities, was countered by the equally plausible points made by **Anne Fairbrother** and **B.L. Lasley**. They showed convincingly that the same processes of diagnosis, testing, and prescription for remediation used in human medicine could be applied toward understanding and assisting stressed ecosystems. A dichotomy became apparent between the maintenance of natural systems and processes and the use of ecosystems for human purposes, but compromise developed around the idea of seeking to maintain the natural trajectories of ecosystems rather than attempting to hold tenaciously to a static but familiar environment.

Pierre Mineau's rhetorical question, "Will we *ever* understand stressed ecosystems enough to regulate the stressors?" relates to this subject directly. The attempts to rehabilitate damaged ecosystems must not exclude legitimate and carefully controlled uses of them. Even with incomplete understanding, it should be possible to prioritize chemical threats and put them under some sort of control. While we do need to apply existing and future knowledge of ecosystem properties and processes increasingly in the construction of environmental policy, D.W. Anderson among others reminded us about ecosystem resiliency: "One individual's catastrophe is another's opportunity."

In the long run, "ecosystem catastrophe" might even be seen as an oxymoron. As Dr. Fairbrother's illustrations show so well, "catastrophe" basically happens only to individuals; ecosystems persist. Measures already available to determine at least the short-term health or sustainability of many ecosystem components include a range of biomarkers, detailed by Mineau and others, and emerging computer models and new physical and chemical measurements. We are witnessing the appearance of three-dimensional computer models such as those described and used by Dr. Adams, and the novel approaches to ecological risk assessment discussed by **T.E. McKone**. Obviously, the development of these and other powerful methods must continue if we are to gain better understanding, protection, and proper utilization of ecosystems. And, as Dr. Goldman pointed out, "Careful monitoring can be good research," and improved monitoring techniques such as **B. D. Hammock**'s immunoassays already are becoming widely available.

On the more biological side, **R. S. Tjeerdema**'s elegant and noninvasive biochemical measurements in aquatic animals by *in vivo* NMR may be extendable to more complex systems, perhaps eventually including simple ecosystems, and physiological endpoints such as those described by Dr. Heath can be applied widely to ecosystem components. Many scientists are just beginning to realize the utility of invertebrate larvae in research on chemical stressors and, of course, their basic importance to aquatic food chains as demonstrated by **G. N. Cherr**. As primary producers, algae, too, deserve far more ecotoxicological attention.

Despite the disadvantages detailed here, the biomarkers described by many of the authors have a bright future in ecotoxicological research and application. However, verity lies in the investigation of *real* populations, communities, and ecosystem processes as they exist in the field. The wedding of field and laboratory methods will be essential for success, as pointed out by both **J. N. Seiber** and Bill Lasley.

It seems apparent that greatly increased teamwork and joint effort will be required in order to accomplish all this. For too long, chemists, toxicologists, and ecologists have talked primarily to one another and in their own tongues. Ecotoxicology seems to be bringing these rather disparate disciplines closer together and introducing a common language. Now it becomes essential that this language be translated for the public and their government representatives, lest even the most substantial of our research efforts be ignored in the construction of policy.

This volume, and the conference on which it is based, have taken a step in that direction. One of our distinguished conference panelists, Dr. Goldberg, remarked that "this is not a 20th-century conference; it belongs in the 21st century," and I agree. Thanks must go to all of the authors and conference panelists, to my fellow organizers, and to our University of California sponsors — the Ecotoxicology Program and the Center for Ecological Health Research on the Davis campus — for contributing to its success.

Index

C

D

E

G

H

I

K

L

M

R

S

T